HTC Vive Focus Vision User Guide

Exploring Virtual Reality Features
and Capabilities

Linda B. Jordan

Copyright

All rights reserved. No part of this publication may be reproduced,
distributed, or transmitted in any form or by any means, including
photocopying, recording, or other electronic or mechanical methods,
without the prior written permission of the publisher, except in the case
of brief quotations embodied in critical reviews and certain other
noncommercial uses permitted by copyright law.
Copyright © (Linda B. Jordan),   (2024).

Table of contents

1. Introduction
Overview of HTC Vive Focus Vision
Key Features and Benefits
Package Contents

2. Getting Started
Unboxing the HTC Vive Focus Vision
Initial Setup and Charging
Installing the Vive Focus Vision App
Connecting to Wi-Fi

3. Hardware Overview
Headset Components and Controls
Front-Facing Tracking Cameras and Sensors
Eye-Tracking Functionality
Lens Adjustment and Comfort Features
Controllers Overview
Connecting Accessories (Facial Tracker, Ultimate Trackers)

4. Headset Setup
Adjusting the Headset for Comfort
IPD (Interpupillary Distance) and Eye Relief Adjustment
Fitting the Strap and Battery Placement
Turning the Headset On/Off
Adjusting the Lenses for Optimal Viewing

5. Using the Vive Focus Vision Standalone Mode
Navigating the Interface
Accessing the Vive Store
Downloading and Installing
Using Mixed Reality (MR) Features
Using Color Passthrough

6. Using the Vive Focus Vision for PC VR
Setting Up the DisplayPort Adapter
Connecting to PC VR via DisplayPort
Configuring SteamVR and Viveport for PC VR
Wireless and Wired Streaming Options
Troubleshooting Common PC VR Issues

7. Eye Tracking & Foveated Rendering
Understanding Eye Tracking
Calibrating Eye Tracking
How Eye Tracking Improves VR Experience
Setting Up Dynamic Foveated Rendering

8. Tracking and Movement
Hand Tracking Setup and Calibration
Using the Vive Focus Vision Controllers
Tracking Modes and Sensor Calibration
Using Body Tracking with Additional Accessories

9. Maintenance & Care
Cleaning and Storing the Headset
Protecting the Lenses and Display
Firmware and Software Updates

Troubleshooting Common Issues
Battery Care and Charging Tips

10. Advanced Features
Customizing the User Interface
Multi-User Setup and Profiles
Using Face and Body Tracking (with Add-ons)
Using the Vive Wired Streaming Kit

11. VR Gaming and Applications
Installing and Playing VR Games
How to Use Viveport and SteamVR
Recommended Apps and Games
Playing VRChat and Other Social Apps

12. Connecting with Other Devices
Pairing with Bluetooth Devices (Headphones, Controllers)
Using the Vive Focus Vision with Other VR Accessories
Casting VR Content to Other Screens

13. Troubleshooting and FAQs
Common Setup Issues and Solutions
Resolving Display and Performance Problems
Eye Tracking Not Working Properly
Connectivity Problems with PC VR
DisplayPort Issues

14. Safety & Health
VR Safety Tips for Comfortable Use
Reducing Eye Strain and Motion Sickness
Proper Posture and Movement Guidelines

How to Take Breaks and Stay Safe

15. Technical Specifications
Display and Lens Specifications
Camera and Sensor Details
Processor and Memory Information
Battery Life and Charging Time
Supported Video and Audio Formats

16. Warranty and Support
Warranty Coverage and Terms
Getting Technical Support
Registering Your Product
HTC Customer Service Information

17. Conclusion
Summary of Features
Tips for Optimizing Your VR Experience
Contact Information for Additional Help

Introduction

The HTC Vive Focus Vision is the latest in HTC's lineup of immersive virtual reality (VR) and mixed reality (MR) headsets. It represents a significant step forward from its predecessor, the Vive Focus 3, by incorporating several innovative features designed to appeal to both consumers and businesses. Unlike traditional VR headsets, which rely solely on virtual reality experiences, the Vive Focus Vision introduces advanced mixed reality capabilities, bridging the gap between the virtual world and the real one.

This standalone headset offers a blend of high-quality VR immersion with the ability to interact with the real world, allowing users to see their physical surroundings while still being deeply involved in their virtual environments. The Vive Focus Vision is powered by the Snapdragon XR2 Gen 1 chipset, known for its ability to handle both VR and AR (augmented reality) applications, ensuring a seamless experience. Additionally, it supports color passthrough, which lets users experience the mixed reality features with full-color, real-time video of the world around them. The headset's eye-tracking technology allows for dynamic adjustment, further enhancing realism and reducing the need for manual configurations like interpupillary distance (IPD) settings.

The Vive Focus Vision is marketed as a versatile, all-in-one solution for anyone looking for high-end, immersive VR experiences, including gaming, simulation, professional applications, and more. Whether used for business simulations, training environments, or leisure activities like gaming and social interactions in virtual spaces, the Vive Focus Vision offers a compelling mix of features designed to appeal to a broad audience.

Key Features and Benefits:

The HTC Vive Focus Vision comes packed with several standout features that make it a versatile and high-performing device, ideal for both consumers and enterprise users. Below are some of the key features and their associated benefits:

1. Color Passthrough & Mixed Reality:

One of the key selling points of the Vive Focus Vision is its color passthrough feature. This allows users to experience mixed reality, seeing their physical environment in full color while interacting with virtual objects. This feature is especially useful for tasks that require awareness of the physical space around you, such as setting up virtual environments or collaborating in mixed-reality spaces. This color passthrough works with the dual cameras and depth sensors at the front of the headset, providing an immersive and seamless blend of real and virtual worlds.

2. Eye Tracking Technology:

Eye tracking is another major feature of the Vive Focus Vision, offering several benefits to users. Not only does it enable dynamic foveated rendering, which improves performance by reducing the graphical load in peripheral areas of the display, but it also enables automatic adjustments to interpupillary distance (IPD) for more precise alignment with the eyes. This feature enhances comfort, reduces eye strain, and ensures a more realistic and immersive visual experience.

3. High-Resolution Display:

The Vive Focus Vision is equipped with dual 2,448 x 2,448 resolution LCDs that deliver crisp, clear visuals. With a 90Hz refresh rate and a wide 120-degree horizontal field of view, the display provides a broad and immersive visual experience, enhancing the overall realism of the virtual environments.

Whether you're playing VR games or working with professional VR applications, the high-resolution display ensures that details are sharp, reducing the "screen door effect" commonly seen in lower-end VR headsets.

4. Powerful Snapdragon XR2 Gen 1 Chipset:

Powered by the Snapdragon XR2 Gen 1, the Vive Focus Vision offers robust performance that can handle both VR and AR experiences. This chipset provides the necessary power for smooth and responsive performance, ensuring that the headset can run graphically intensive applications without significant lag or issues. The chipset also supports enhanced processing power for the headset's mixed reality capabilities, allowing users to experience seamless transitions between the virtual and physical worlds.

5. Standalone Functionality:

One of the most notable benefits of the Vive Focus Vision is its ability to function standalone without the need for a PC or external sensors. This means users can quickly set up and dive into VR or MR experiences without worrying about cables or additional equipment. The headset features 12GB of RAM and 128GB of storage, providing ample space for apps and content. While it is compatible with PC VR setups for more immersive experiences, it is primarily designed to be an all-in-one, portable VR solution.

6. Comfort and Ergonomics:

The Vive Focus Vision is designed for extended wear, offering a comfortable fit even during long sessions. The headset uses a rigid plastic frame with a well-balanced design that distributes the weight evenly across the head. The battery is located at the rear of the headset, helping to balance the device's weight and reduce strain on the front of the head. Additionally, the lenses are adjustable to ensure a personalized and comfortable fit. These ergonomic features ensure that users can enjoy their VR experiences for extended periods without discomfort.

7. PC VR Compatibility:

While the Vive Focus Vision is a standalone device, it also supports PC VR experiences via an optional DisplayPort adapter. This adapter enables users to connect the headset to a gaming PC for lossless VR experiences, utilizing the power of a high-performance GPU. This feature makes the Vive Focus Vision a flexible device that can meet the needs of both casual users looking for standalone VR and professional users requiring high-fidelity, graphically demanding VR applications.

Package Contents:

When you purchase the HTC Vive Focus Vision, the package includes everything you need to get started with the headset right away. Here's what you can expect to find in the box:

1. Vive Focus Vision Headset:

The main device, featuring the high-resolution display, eye-tracking sensors, front-facing cameras, and depth sensors, as well as the necessary internal components for standalone VR and mixed reality experiences.

2. Controllers:

The package includes the Vive Focus 3 controllers, which are ergonomic and intuitive, designed to provide accurate motion tracking and responsiveness for your hands in VR environments.

3. Power Adapter & USB-C Cable:

A power adapter is included for charging the headset, along with a USB-C cable to connect to the charging port.

4. Vive Wired Streaming Kit (Pre-Order Bonus):

For those who pre-order the Vive Focus Vision, HTC offers a Vive Wired Streaming Kit as a bonus. This kit includes a DisplayPort adapter and USB-A cable, allowing you to connect the headset to your PC for high-quality, lossless PC VR streaming.

5. Comfort Accessories:

The box also contains accessories to ensure a comfortable fit, such as adjustable straps, padding for the head, and optional lens covers for protection.

6. Quick Start Guide:

A user manual to guide you through the setup process, including how to adjust the headset for your comfort, install software, and connect to PC VR if desired.

In conclusion, the HTC Vive Focus Vision is a versatile and powerful headset that brings a blend of immersive VR and mixed reality features into a standalone device. With its high-resolution display, eye-tracking, and PC VR compatibility, it offers users an unmatched experience in both virtual and physical environments. Whether for gaming, professional use, or mixed-reality interactions, the Vive Focus Vision is designed to cater to a wide range of applications, making it a highly anticipated addition to the VR headset market.

Chapter 1.

Getting Started

Unboxing the HTC Vive Focus Vision is an exciting moment for any new user. As you open the sleek, black box, the first impression is one of premium design and careful organization. Inside, everything is meticulously packed to ensure that the device arrives in perfect condition, ready for your first experience with mixed reality.

The box contains the Vive Focus Vision headset, the main device, nestled securely with protective foam inserts to prevent damage during transit. The first thing you'll notice is the sleek, futuristic design of the headset. The lightweight, matte-black finish of the headset offers a high-end feel, and the prominent front-facing cameras and sensors, which give it the capability for mixed-reality experiences, are visible right at the front.

Below the headset, you'll find the controllers, designed to be comfortable and ergonomic for extended use. The controllers are packaged in separate compartments to keep them safe during shipping. These controllers are intuitive, designed to fit in your hands naturally, and feature a variety of buttons and trackpads for controlling your VR or MR experience.

Also included in the package are the power adapter and USB-C charging cable, necessary for powering the headset and keeping it charged. A bonus accessory, the Vive Wired Streaming Kit, may also be included if you pre-ordered the headset. This kit comes with a DisplayPort adapter and USB-A cable for connecting to a PC for high-fidelity VR experiences, ensuring you're ready for both standalone and PC VR use.

There's also a quick-start guide included in the box. This user manual provides helpful instructions on how to get your headset up and running. HTC also includes comfort accessories, such as an adjustable strap and extra padding for the headset to ensure the device fits securely and comfortably on your head.

Once you've unpacked everything, you're ready to start the setup process.

Initial Setup and Charging:

The initial setup of the HTC Vive Focus Vision is straightforward, but there are a few important steps to ensure everything is done properly.

1. Charge the Device:

Before you begin the setup process, it's important to ensure that your Vive Focus Vision has sufficient charge. Plug the USB-C cable into the headset and the included power adapter, and connect it to a power outlet. Charging the device initially is essential, as it ensures that you won't run into any power issues during setup. Typically, the headset needs about 2–3 hours to fully charge, but you can still set up and use it while it's charging, as long as there's enough battery to power the device.

2. Powering On the Headset:

Once the device is charged, power it on by holding the power button located on the side of the headset. The HTC Vive logo will appear on the display as the device boots up. You may also see a welcome screen guiding you through the rest of the setup process.

3. Adjusting the Fit:

After powering on the device, you'll be prompted to adjust the head strap to ensure a snug fit. The Vive Focus Vision is designed to be comfortable for long sessions, and the adjustable straps and face padding help achieve the best fit for your face. Be sure to adjust the lens distance as well, as this ensures your vision is optimized for your eye shape.

4. Initial Calibration:

As you begin the setup, the system will run an initial calibration to optimize the headset's eye-tracking and display settings. The headset will prompt you to look at certain points, ensuring that the device calibrates your interpupillary distance (IPD) and ensures accurate eye tracking. This calibration process helps to improve comfort, reduce eye strain, and ensure the clearest visuals during use.

Once your headset is fitted and calibrated, you're ready to begin exploring the software setup.

Installing the Vive Focus Vision App:

Before diving into virtual experiences, you need to install the Vive Focus Vision app on your mobile device. This app is essential for managing the headset, downloading content, and connecting the device to Wi-Fi.

1. Download the App:

The Vive Focus Vision app is available for download from both the Apple App Store and Google Play Store. Simply search for "Vive Focus Vision" in the app store of your choice and download the app to your smartphone. It's important to use a smartphone with Bluetooth 4.0 or later and Android 7.0 or later or iOS 12.0 or later for optimal compatibility with the Vive Focus Vision.

2. Create or Log into Your HTC Account:

Upon opening the app, you'll be prompted to either create a new HTC account or log into an existing one. Your HTC account is necessary for syncing your device settings, purchasing content from the Viveport store, and managing software updates. If you don't already have an account, you can create one directly through the app with your email address and a password.

3. Pairing the Headset:

Once you've logged in, the next step is to pair your Vive Focus Vision headset with your smartphone. The app will guide you through the pairing process, which typically involves turning on the headset and enabling Bluetooth on your smartphone. After a few moments, the app should detect the headset, and you'll be prompted to pair them. Ensure that the headset is within range of your phone for a successful connection.

4. Software Updates:

Before you begin using the Vive Focus Vision, it's a good idea to check for any software updates that may be available. The app will prompt you if an update is necessary, and installing the latest firmware ensures you have access to new features and optimizations. Updates typically take only a few minutes but should not be interrupted.

Connecting to Wi-Fi:

Once your headset is paired with your mobile device, it's time to connect to Wi-Fi to begin downloading apps, games, and updates. Connecting to Wi-Fi ensures that your Vive Focus Vision is able to access the Viveport store for content and any necessary online services.

1. Access the Wi-Fi Settings:

On the Vive Focus Vision, navigate to the Wi-Fi settings by selecting the Wi-Fi icon in the menu. The device will scan for available networks, and a list of Wi-Fi networks within range will appear on the screen.

2. Select Your Network:

From the list of available networks, select the one you wish to connect to. If your network is secured, you'll be prompted to enter your Wi-Fi password. Enter the password carefully, as the headset may not allow you to proceed without the correct credentials.

3. Confirm the Connection:

After entering the password, the headset will attempt to connect to the Wi-Fi network. You should see a confirmation message once the connection is established. If the headset cannot connect, make sure the Wi-Fi password is correct and that the network is stable.

4. Testing the Connection:

Once connected, you can test the Wi-Fi connection by browsing the Viveport store or streaming content. This will ensure that the headset is properly connected to the internet and that all online services will function smoothly.

Now that the headset is fully set up, you can start exploring the vast world of VR and MR experiences. With the initial setup completed, the Vive Focus Vision is ready to deliver high-quality immersive content with ease. Whether you're diving into games, exploring virtual environments, or engaging in mixed-reality applications, your HTC Vive Focus Vision is ready to take you on a journey into the future of digital interaction.

Chapter 2.

Hardware Overview

The HTC Vive Focus Vision headset is a marvel of modern design, combining cutting-edge technology with a user-friendly form factor. The headset is designed to offer an immersive experience while remaining comfortable and easy to use, even during extended sessions.

The main component of the headset is the display unit, which includes dual 2,448 x 2,448 resolution LCD screens, one for each eye. These high-resolution displays provide clear and vibrant visuals, ensuring that you can enjoy immersive mixed-reality (MR) experiences with minimal pixelation. The screens are encased in a durable yet lightweight housing, with ventilation holes to prevent overheating during extended use. The front-facing part of the headset also houses Fresnel optics, which help to reduce glare and enhance contrast for a more comfortable viewing experience.

The head strap of the Vive Focus Vision is designed for both comfort and adjustability. It features a dial system at the back, which allows users to easily adjust the tightness for a secure fit. The head strap is made of soft materials that are comfortable against the skin, and the face padding is soft and breathable to help keep the user cool during long VR sessions. The adjustable design ensures that users of different head shapes and sizes can enjoy a snug fit, enhancing both comfort and stability.

On the sides of the headset, you'll find physical buttons that are used to control the power, volume, and other basic functions of the headset. The power button is located on the right side of the device, and the volume buttons are positioned on the left. These controls are intuitively placed, ensuring that users can adjust settings without needing to remove the headset.

In addition to these physical controls, the Vive Focus Vision has touch-sensitive areas on the sides that are used for navigation and interaction within the VR environment. These touchpads allow for easy scrolling, zooming, and selecting options, offering an intuitive and natural way to navigate the interface.

Front-Facing Tracking Cameras and Sensors:

One of the standout features of the HTC Vive Focus Vision headset is its advanced front-facing tracking cameras and sensors, which are key to providing an immersive mixed-reality (MR) experience. These cameras and sensors allow the headset to map the physical environment around you, enabling it to seamlessly blend virtual and real-world elements. This capability is at the heart of mixed reality, allowing you to see your surroundings while interacting with virtual objects.

There are two color cameras on the front of the headset that capture the scene in real time, while a pair of depth sensors map out the distance between objects in the physical space. This combination of cameras and sensors works in tandem to create a real-time representation of the environment, which is then overlaid with virtual content. This allows you to interact with both the virtual and real world simultaneously, offering a more immersive and natural experience than traditional VR.

In addition to capturing the environment, these cameras enable color passthrough functionality. This allows you to view the real world in color, rather than the black-and-white world that most VR headsets offer. This makes it easier to navigate your physical environment while wearing the headset, whether you're setting up your play area or simply checking your surroundings. The depth sensors also provide enhanced tracking of physical objects, ensuring that your virtual content behaves naturally when interacting with the real world.

This advanced camera system also enables features such as hand tracking and room-scale tracking, where the headset can accurately track your hand movements and adjust the virtual environment based on your physical location. These features make the Vive Focus Vision not only a top-tier VR device but also a powerful tool for mixed-reality experiences.

Eye-Tracking Functionality:

The eye-tracking functionality in the HTC Vive Focus Vision is a breakthrough feature that significantly enhances both the user experience and the performance of the device. The integrated infrared sensors and cameras within the headset track the movement of your eyes, enabling the system to understand exactly where you're looking at any given time. This functionality has multiple benefits that elevate the overall experience.

1. Foveated Rendering:

Eye-tracking allows for foveated rendering, a technology that reduces the graphical load on the device by rendering only the area where the user is directly looking in full resolution. Areas outside the user's focal point are rendered at a lower resolution, improving performance and efficiency without sacrificing visual quality in the areas that matter most. This is particularly useful for maintaining high frame rates and reducing latency in virtual environments.

2. Gaze-Based Interactions:

Eye-tracking also enables gaze-based interactions. For instance, the system can detect where you are looking and allow you to interact with virtual objects by simply focusing your gaze on them for a certain period of time. This is especially useful in mixed-reality applications, where precision is key to interacting with virtual content in a natural way.

3. Personalized Comfort:

The eye-tracking system can be used to adjust the headset's display settings for optimal comfort. It can automatically adjust the IPD (interpupillary distance) for a more comfortable and customized fit, reducing strain and improving clarity.

Lens Adjustment and Comfort Features:

The HTC Vive Focus Vision headset comes with a number of lens adjustment options that ensure a comfortable and personalized viewing experience. The lenses are adjustable to cater to different users' needs, particularly when it comes to the interpupillary distance (IPD). The IPD refers to the distance between your eyes, which varies from person to person. Proper IPD adjustment ensures that the images displayed by the headset are clear and properly aligned with your eyes, reducing eye strain and improving immersion.

The lens adjustment knob on the headset allows you to fine-tune the distance between the lenses, ensuring the clearest possible image. This is especially important for users who may have a wide or narrow IPD, as it helps prevent blurry visuals and ensures a sharp, focused image.

Comfort features extend beyond the lens adjustments as well. The face padding is made from soft materials that are designed to be breathable, reducing sweat and discomfort during extended VR sessions. Additionally, the adjustable head strap ensures a snug fit, preventing the headset from shifting during movement. The combination of these comfort features makes the Vive Focus Vision comfortable for extended periods, even for users who may wear glasses.

Controllers Overview:

The HTC Vive Focus Vision comes with a pair of controllers that are designed to offer a natural and intuitive way to interact with the virtual environment. These controllers are equipped with trackpads and buttons that enable precise navigation and interaction with virtual elements. The trackpads are responsive to both touch and swipe gestures, making them ideal for navigating menus, selecting items, and controlling movement within VR experiences.

In addition to the trackpads, the controllers have trigger buttons for actions such as grabbing or interacting with objects, as well as grip buttons that allow for more nuanced interactions. These buttons are designed to be easily accessible and responsive, making the controllers feel like an extension of your own hands.

The controllers are also equipped with motion sensors and gyroscopes, which enable accurate tracking of hand movements in space. This provides a more immersive experience, as the virtual world reacts in real time to the movements and actions of the user.

Connecting Accessories (Facial Tracker, Ultimate Trackers):

To enhance the mixed-reality experience, the HTC Vive Focus Vision is compatible with several accessories, including the Vive Facial Tracker and Ultimate Trackers. The Vive Facial Tracker is an optional accessory that attaches to the front of the headset and provides enhanced facial expression tracking. This feature allows for more immersive and realistic virtual interactions, as your facial expressions are mirrored by your digital avatar. This is especially useful for applications that require social interaction or communication, such as multiplayer VR games or virtual meetings.

The Ultimate Trackers are designed to work with the Vive Focus Vision to provide more precise full-body tracking. These trackers can be attached to your body, such as your feet, waist, or wrists, and they communicate with the headset to provide accurate movement data. This enables a more immersive experience, especially in VR applications that involve complex movement or physical interaction.

These accessories are optional but offer significant improvements in the overall VR experience, providing users with more accurate tracking and a deeper sense of presence within virtual worlds.

With these hardware components and features, the HTC Vive Focus Vision stands out as a powerful and versatile mixed-reality headset, designed to deliver a rich, immersive experience that can be customized and fine-tuned for each user. Whether you're exploring virtual environments, engaging in mixed-reality applications, or using the headset for productivity, these features work together to enhance your experience.

Chapter 3.

Headset Setup

Achieving the perfect fit and comfort with your HTC Vive Focus Vision headset is essential for enjoying long VR sessions without discomfort. The device is engineered to provide a customizable fit, ensuring that the headset stays securely in place and feels comfortable even after hours of use. The key elements to adjust for comfort include the head strap, lens placement, and face padding.

The head strap is designed to be easily adjustable, and its comfort is critical for providing a secure and snug fit without placing unnecessary pressure on your head. Start by loosening the strap and placing the headset on your head. Then, use the adjustment dial located at the back of the head strap to fine-tune the fit. The dial allows for quick adjustments to tighten or loosen the strap, ensuring that the headset stays firmly in place without causing discomfort.

Once the headset is secured on your head, it's important to check the face padding. The padding on the inside of the headset serves not only to provide comfort but also to help block out light and improve the immersive experience. Make sure the padding is positioned correctly on your face, with the edges of the foam gently resting against your skin. This will prevent light leaks that can disrupt your VR experience and ensure a comfortable fit for extended sessions.

Finally, ensure that the headset feels balanced. A poorly balanced headset can lead to strain on your neck and discomfort. If the weight feels uneven, adjust the head strap slightly to better distribute the weight across your head.

IPD (Interpupillary Distance) and Eye Relief Adjustment:

One of the key factors in ensuring a clear, comfortable view during VR sessions is proper Interpupillary Distance (IPD) adjustment. IPD refers to the distance between the center of your pupils, and getting this setting right is crucial for avoiding eye strain and ensuring the clearest image. A headset like the HTC Vive Focus Vision allows for manual IPD adjustment to cater to individual users' needs.

The IPD adjustment mechanism on the Vive Focus Vision is located on the bottom of the headset. It is typically controlled via a dial or slider that enables you to adjust the distance between the lenses. Start by adjusting the IPD so that the two lenses are spaced to match your natural eye distance. To achieve the best clarity, adjust the IPD until the virtual image looks crisp and aligned.

When adjusting the IPD, it's important to focus on getting a clear view in the center of your field of vision. This will minimize distortion around the edges of your view, ensuring a more natural and immersive experience. If you wear glasses, the correct IPD setting can also prevent discomfort by ensuring that the lenses don't push against your eyewear.

In addition to IPD, eye relief is another adjustment that may be important for some users. Eye relief refers to the distance between the lenses and your eyes, and it can be adjusted to avoid any discomfort, especially for those who wear glasses. The Vive Focus Vision headset provides a fixed distance for the lenses, but adjusting how the headset sits on your face by manipulating the head strap can help you achieve optimal eye relief. Ensuring your eyes are properly aligned with the lenses is important for minimizing glare and improving image clarity.

Fitting the Strap and Battery Placement:

Proper placement and adjustment of the head strap are essential to ensure that the headset remains comfortable and securely in place during use. After adjusting the head strap's tightness, it's also important to consider the placement of the battery unit, which is located at the back of the headset. The battery plays a critical role in powering the headset and contributes to the overall weight distribution.

The Vive Focus Vision features an integrated battery pack at the rear of the headset, which is designed to help balance the weight of the device. The battery itself should rest comfortably at the back of your head. Ensure that the battery is aligned with the back of your skull to prevent discomfort and unnecessary strain on your neck. When the battery is properly positioned, it helps distribute the weight evenly across your head, preventing any uncomfortable pressure points.

The head strap's design also includes a cushioned area where the battery rests, which ensures comfort during long sessions. It's important to adjust the strap in such a way that the battery unit sits snugly against the back of your head, but not too tight to cause discomfort. If you feel that the strap is too loose or too tight, make adjustments until the fit feels comfortable, with the battery contributing to an even distribution of weight.

Once the head strap and battery are properly adjusted, it's important to check that the headset does not shift during movement. If the headset feels unbalanced or loose, you may need to readjust the strap or reposition the battery to achieve a better fit.

Turning the Headset On/Off:

Turning the HTC Vive Focus Vision headset on and off is a simple process, but it's essential to do it correctly to ensure that the device powers up and shuts down safely. The power button is located on the right side of the headset, and it is designed for easy access even when the headset is worn.

To turn on the headset, press and hold the power button for a few seconds until you see the Vive logo appear on the screen. You should also hear a sound indicating that the headset is powered on. Once the device is on, it will enter its startup sequence, which may involve loading system software and connecting to any paired devices or Wi-Fi networks.

To turn off the headset, press and hold the power button again for a few seconds. The screen will prompt you with a confirmation message, allowing you to power off the device completely. If the headset is not being used for an extended period, it's best to power it off to conserve the battery.

To ensure the headset is always ready for use, it's recommended to charge the Vive Focus Vision regularly. Charging is done via a USB-C charging port located on the side of the device. The headset comes with a dedicated charging cable and power adapter, which you can use to charge the device either through a wall outlet or via a computer. When the battery reaches full charge, the LED indicator on the headset will change from red to green, signaling that it's ready to use.

Adjusting the Lenses for Optimal Viewing:

The lens adjustment in the HTC Vive Focus Vision headset is essential for achieving the clearest, sharpest visual experience. The Vive Focus Vision allows you to fine-tune the lenses to suit your specific visual needs, enhancing your VR experience and reducing eye strain.

While the device does not offer a physical lens adjustment for diopters (like some other VR headsets), it does offer IPD (Interpupillary Distance) adjustments, which will directly affect how the lenses align with your eyes for the best view. Ensure that the IPD setting is aligned with your natural eye distance, and use the adjustment dial to tweak the alignment until the image is sharp and clear.

When adjusting the lenses, it's important to position the headset correctly on your face to ensure that the lenses are in the optimal location for your eyes. Avoid having the lenses too close to or too far from your eyes, as this can cause blurriness, distortion, or discomfort. For the best results, make sure the lenses are aligned with the center of your eyes and ensure that the image is clear with minimal distortion around the edges.

In conclusion, adjusting the HTC Vive Focus Vision headset to ensure comfort, clarity, and proper alignment is a critical step to achieving the best possible virtual reality experience.

By taking the time to fine-tune the fit, IPD, and lens alignment, you can ensure a comfortable and immersive VR experience that minimizes discomfort and maximizes enjoyment.

Chapter 4.

Using the Vive Focus Vision Standalone Mode

The HTC Vive Focus Vision is a powerful mixed-reality (MR) and virtual reality (VR) headset that combines immersive experiences with practical functionality. Understanding how to fully utilize the features and interface of the Vive Focus Vision enhances your overall experience, whether you're engaging in virtual environments or exploring mixed-reality content. This chapter will guide you through the Vive Focus Vision's standalone mode, interface navigation, accessing the Vive Store, downloading and installing apps, and using mixed reality features.

Standalone Mode:

One of the standout features of the HTC Vive Focus Vision is its standalone mode, which means that the device doesn't require an external PC or console to operate. This offers unparalleled convenience for users, as you can enjoy VR and

MR experiences without being tethered to any wires or additional hardware. The Vive Focus Vision is equipped with a powerful chipset, the Snapdragon XR2 Gen 1, which provides the necessary computational power to run a wide range of virtual and mixed-reality applications on its own.

To use the Vive Focus Vision in standalone mode, you simply need to power on the headset, adjust the fit for comfort, and connect to Wi-Fi for app downloads and updates. With the absence of any external device, everything from immersive VR games to mixed-reality applications runs seamlessly from the internal hardware. This makes the Vive Focus Vision an excellent choice for users who want portability and ease of use while still accessing high-quality VR experiences.

In standalone mode, the Vive Focus Vision is capable of running both VR games and MR apps that are compatible with the system, using the internal storage and processing power. You can also access a variety of cloud-based services that will allow you to store and sync data across devices, as well as stream content directly to the headset.

Navigating the Interface:

Navigating the HTC Vive Focus Vision interface is intuitive, thanks to its user-friendly layout. The interface is designed to be easily accessible whether you're in VR mode or mixed-reality mode. Upon powering on the headset, you are greeted with a welcome screen that provides easy-to-follow instructions for first-time setup, as well as access to key features and apps.

The interface consists of a home menu, from which you can launch apps, access settings, and view notifications. This menu is navigated using the controller, which operates via point-and-click functionality. You can scroll through the available options, select items, and make adjustments by pointing to the desired icon and clicking the controller's trigger button.

The Vive Focus Vision interface also provides a quick access menu for common actions such as adjusting volume, screen brightness, Wi-Fi settings, and more. To open the quick access menu, you simply press the button on the controller. This provides instant control over some of the more important aspects of the headset's functioning without having to navigate through multiple menus.

To enhance ease of use, the interface includes the ability to customize certain elements. For instance, you can rearrange the apps on your home screen to prioritize the ones you use the most, or change the visual layout to make it easier to navigate.

Accessing the Vive Store:

To make the most out of your HTC Vive Focus Vision experience, you'll need to access the Vive Store the official marketplace where you can browse, purchase, and download VR and MR applications and games. The Vive Store offers a wide variety of content, ranging from entertainment and gaming to educational and professional applications, ensuring that there's something for everyone.

Accessing the Vive Store is simple. Once you're inside the main menu, you'll find a prominent option for the Vive Store icon. Selecting it will bring you to the store interface, where you can browse the available apps by category or use the search function to find specific titles. The Vive Store features both free and paid content, so there's plenty of options for those looking to explore without spending money upfront.

The store interface is clean and well-organized. The front page often showcases popular apps, new releases, and special deals or discounts. You can scroll through the available content and read detailed descriptions, reviews, and ratings for each app to help you decide which ones to try. Once you've made your selection, you can purchase or download an app directly to your device. Depending on your preferences, you can enable automatic updates for apps to ensure you always have the latest features and fixes.

Additionally, the Vive Store allows you to sync your purchases across devices, so you can access any app you've bought or downloaded on other compatible Vive headsets if needed.

Downloading and Installing Apps:

The process of downloading and installing apps on the Vive Focus Vision is straightforward and user-friendly. Once you've located an app in the Vive Store that you want to try, simply click on it to view more details. If the app is free, you can click Download and it will begin the installation process automatically. For paid apps, you'll be prompted to complete the purchase before the download begins.

Once the download begins, the installation process will automatically start in the background. You'll be able to see the progress on the screen as the app is being installed. When the installation is complete, the app will appear in your library, ready to be launched.

After installation, you can access the app from the home screen or your app library. If you ever wish to remove an app, you can simply press and hold its icon on the home screen, and an option to uninstall the app will appear.

The Vive Focus Vision supports cloud storage for many apps, allowing you to back up your data and sync it across devices. This is particularly useful for apps that require high storage or save game progress. This means you can install an app, use it for a while, and if you need to free up space, uninstall it knowing that your progress will be saved in the cloud and can be restored when you reinstall the app later.

Using Mixed Reality (MR) Features:

One of the most exciting features of the HTC Vive Focus Vision is its ability to deliver mixed-reality (MR) experiences. Mixed reality blends virtual environments with the real world, allowing users to interact with both at the same time. This offers a more immersive and interactive experience compared to traditional VR.

The Vive Focus Vision uses color passthrough cameras to enable mixed-reality capabilities. This means that, while wearing the headset, you can see the real world in color through the camera lenses, allowing you to interact with your physical environment while also experiencing virtual elements.

To access MR features, you can enter Mixed Reality Mode by selecting the appropriate setting from the main menu or activating it within supported apps. In this mode, the real-world environment is integrated into your virtual space, and the headset tracks your movements to interact with both physical and virtual objects. For example, you can play games where virtual objects are placed in the real world, or use MR applications that let you visualize 3D models in your physical surroundings.

The ability to toggle between MR and VR modes provides flexibility depending on the type of experience you want to have. Whether you're sitting comfortably in a chair for a VR experience or moving around the room for an MR experience, the Vive Focus Vision adapts to your preferences.

Using Color Passthrough:

Color passthrough is one of the key features that enables the Vive Focus Vision to offer high-quality mixed-reality experiences. This feature allows you to see your physical environment in full color while still wearing the headset. The dual front-facing cameras are responsible for capturing the real world around you, and the passthrough feature ensures that the visual feed is displayed on the internal screen.

This capability is especially important when navigating physical spaces while wearing the headset. Whether you're adjusting the position of virtual objects or simply want to be aware of your surroundings, color passthrough helps you safely interact with the physical world without removing the headset.

The passthrough feature can be activated through the menu settings or by pressing a button on the controller. You can adjust the opacity and transparency of the passthrough view to suit your needs, allowing you to control how much of the physical world you see while immersed in virtual content.

In conclusion, the HTC Vive Focus Vision offers a seamless experience for both virtual and mixed-reality users. Its standalone mode ensures portability, while intuitive navigation, access to the Vive Store, and powerful MR features bring an elevated level of immersion. Whether you're playing games, exploring virtual worlds, or interacting with mixed-reality environments, the Vive Focus Vision provides an advanced yet easy-to-use platform for a wide range of immersive experiences.

Chapter 5.

Using the Vive Focus Vision for PC VR

While the HTC Vive Focus Vision is a powerful standalone mixed-reality (MR) headset, one of its key features is the ability to also connect to a PC for even more immersive experiences. By connecting to your PC, you can access a wide range of high-performance VR content from platforms like SteamVR and Viveport, offering a seamless transition from standalone use to high-powered VR. This chapter will guide you through the process of setting up the Vive Focus Vision for PC VR, covering DisplayPort connections, configuration, streaming options, and troubleshooting common issues.

Setting Up the DisplayPort Adapter:

To connect the Vive Focus Vision to your PC for VR gaming or applications, you'll need to set up the DisplayPort adapter. The Vive Focus Vision doesn't come with a built-in wired PC connection by default, but HTC provides a Vive Wired

Streaming Kit to enable this connection. This kit includes an adapter and the necessary cables to connect the headset to your PC's DisplayPort.

Start by plugging the DisplayPort cable into the DisplayPort output of your PC's graphics card. Depending on your PC's configuration, you may need to use an HDMI-to-DisplayPort adapter if your GPU doesn't have a native DisplayPort connection. Afterward, connect the other end of the cable to the Vive Wired Streaming Kit's DisplayPort adapter.

The Vive Focus Vision headset itself connects to the adapter via a USB-C cable. This cable transmits both data (for tracking, interaction, and communication) and power to the headset. Once all connections are made, ensure the headset is properly seated and all cables are securely plugged in to avoid interruptions during gameplay.

Finally, if you haven't already done so, install the Vive software on your PC, which is crucial for the headset to interface with your system and launch compatible applications. You can download the software from the HTC Vive website or access it via SteamVR for a more integrated experience.

Connecting to PC VR via DisplayPort:

Once the DisplayPort adapter is set up, the next step is connecting your Vive Focus Vision to your PC for a PC VR experience. The process of establishing this connection is straightforward but requires ensuring that both your PC and the headset are correctly configured to communicate with each other.

1. Power On the Headset:

Start by turning on the Vive Focus Vision headset using the power button located on the side or top of the device.

2. Launch Vive Software on PC:

On your PC, open the Vive Software or SteamVR application. You will need the Viveport or SteamVR environment to configure and run PC VR apps.

3. Connect the Headset to the PC:

If all the necessary connections are in place, the software should recognize the headset once it's powered on and connected via the DisplayPort adapter. Follow any on-screen prompts on both the headset and your PC to complete the setup.

4. Calibration and Detection:

After the software detects the headset, you'll likely need to go through a brief calibration process to adjust tracking and make sure that the Vive Focus Vision is properly synced with the PC's graphics output. This may involve adjusting settings for frame rate, resolution, and refresh rates to ensure optimal performance during your VR experience.

Configuring SteamVR and Viveport for PC VR:

To get the most out of your PC VR experience, you'll need to set up both SteamVR and Viveport (HTC's proprietary app store) to access compatible VR content.

1. SteamVR Configuration:

First, ensure you have SteamVR installed on your PC. SteamVR is the main platform for VR games on Steam, and it's fully compatible with the Vive Focus Vision.

Open SteamVR and follow the on-screen instructions for setting up your VR hardware. This process will guide you through tracking setup, room-scale configuration, and any initial calibration of the headset.

SteamVR will automatically detect the Vive Focus Vision once it's connected via DisplayPort. Make sure to update SteamVR and all VR apps to the latest versions to ensure compatibility with the headset.

2. Viveport Configuration:

If you prefer to use HTC's Viveport store for your VR content, open the Viveport software on your PC after connecting the headset. Viveport offers both subscription-based and individual purchases for VR content.

Log in to your HTC account, browse for available apps, and install them directly from the Viveport library. Ensure the correct VR mode is selected for the Vive Focus Vision when running content from Viveport, as this will ensure the content is optimized for your headset's display.

Both SteamVR and Viveport offer cross-platform VR content, but some apps may be exclusive to one platform, so it's important to explore both stores for the best selection.

Wireless and Wired Streaming Options:

HTC provides two methods for streaming PC VR content to the Vive Focus Vision: wired streaming and wireless streaming. Each has its advantages depending on your needs.

1. Wired Streaming (Using the DisplayPort Adapter):

Wired streaming offers the best performance and consistency in terms of frame rate, resolution, and latency. By using the DisplayPort adapter, the Vive Focus Vision benefits from a direct connection to your PC's GPU, ensuring high-quality rendering without interruptions.

Advantages:

Better graphics fidelity, no risk of network interruptions, and stable performance.

Best for:

High-intensity VR games or professional applications where visual fidelity and performance are critical.

2. Wireless Streaming (Vive Streaming Kit):

The Vive Focus Vision also supports wireless streaming for those who prefer more freedom of movement. By using the Vive Wireless Adapter, you can stream PC VR content directly to the Vive Focus Vision over Wi-Fi, removing the need for cables.

Advantages:

Enhanced freedom of movement, more immersive experience, and cable-free setup.

Best for:

Casual VR experiences, movement-based apps, or when you need to quickly set up without worrying about cables.

However, wireless streaming may have slightly higher latency or lower frame rates, depending on the quality of your Wi-Fi connection. It's recommended to use Wi-Fi 6 for optimal performance, as it offers the fastest data transfer speeds and the least latency.

Troubleshooting Common PC VR Issues:

While the Vive Focus Vision offers seamless performance when connected to a PC, users may occasionally encounter issues that disrupt their VR experience. Here are some common troubleshooting tips:

1. No Display or Connection Issue:

Ensure all cables are securely connected, including the DisplayPort and USB-C cables.

Confirm that your PC graphics card supports VR and that the latest drivers for both the GPU and the Vive Focus Vision are installed.

Restart both the headset and the PC to reset any connection issues.

2. Latency or Lag in VR:

Check your PC's graphics settings and ensure they're set to high performance for VR content. Lowering graphical settings can sometimes reduce latency.

For wireless streaming, ensure your Wi-Fi connection is stable and that you're using a 5GHz network to avoid interference.

3. Tracking Issues:

Ensure that the Vive Focus Vision cameras have a clear, unobstructed view of the environment.

Recalibrate the room-scale setup or chaperone system in SteamVR to optimize tracking.

4. Audio Issues:

Check that the headset's audio settings are correctly configured and that the sound is enabled in both the PC's system settings and within SteamVR or Viveport.

If using wireless streaming, ensure that the Bluetooth connection is not interfering with audio output.

In conclusion, connecting the HTC Vive Focus Vision to a PC for VR content offers an entirely new dimension of immersive experiences. With both wired and wireless streaming options, configuring SteamVR and Viveport ensures you have access to a wealth of content, while troubleshooting common issues ensures you can maintain a smooth and enjoyable experience. Whether you're playing demanding VR games or exploring mixed-reality environments, the Vive Focus Vision gives you the flexibility and power to enjoy PC VR to its fullest.

Chapter 6.

Eye Tracking & Foveated Rendering

The HTC Vive Focus Vision is equipped with cutting-edge features to enhance the overall virtual reality experience, one of the most prominent being eye tracking and its integration with foveated rendering. These two technologies work together to optimize the performance, realism, and immersion of VR environments, making it possible for users to experience high-quality VR even with limited computing power. This chapter will explore how these features function, how to calibrate eye tracking, and how they work in tandem to provide a more efficient and realistic experience.

Understanding Eye Tracking:

Eye tracking is the process of monitoring and analyzing the movement and position of a user's eyes. In the context of virtual reality, it allows the system to detect where the user is looking at any given moment. The HTC Vive Focus Vision features built-in eye-tracking sensors that can track the movement of both eyes in real time. This technology is essential for several aspects of the VR experience, from interactive elements to enhancing graphical performance.

In the case of the Vive Focus Vision, eye tracking is accomplished with the help of specialized infrared sensors placed around the lenses of the headset. These sensors emit infrared light and measure how it reflects off the user's pupils. Using this data, the system can accurately determine the gaze direction, as well as factors like pupil dilation, which can provide additional information about the user's focus and emotional response during interactions.

This data allows the VR system to deliver more natural and dynamic experiences. Eye tracking can also be utilized to improve interactive elements within a virtual environment, where objects may react based on where the user looks. It also plays a vital role in enabling foveated rendering (explained below), which further elevates the realism and performance of the VR experience.

Calibrating Eye Tracking:

To make sure the eye tracking system is functioning accurately and efficiently, it's important to calibrate the system before using it. Calibration ensures that the Vive Focus Vision can precisely track your eye movements, resulting in a better experience when interacting with the virtual environment. Calibration is a relatively quick and easy process that can be done in just a few steps.

1. Starting Calibration:

When you first set up the headset, the system will prompt you to perform an eye tracking calibration. If you're already using the device and need to recalibrate, you can access the calibration tool from the settings menu in the VR interface.

2. Positioning the Headset:

Before starting calibration, ensure that the headset is properly adjusted on your head. The Vive Focus Vision should be centered over your eyes, with the lenses aligned with your pupils. This step is important because the eye tracking system works best when the headset is in the correct position, as it needs an unobstructed view of your eyes.

3. Calibration Process:

During calibration, the system will ask you to focus on several on-screen points (usually represented as dots or circles) that appear at different locations in the field of view. As you focus on each point, the eye tracking sensors will capture the position of your gaze and refine the system's accuracy. Calibration might take a minute or two, and the system will let you know when it's completed successfully.

4. Adjustments for Comfort:

If you wear glasses or have specific vision needs, you may need to adjust the headset's lens distance or IPD (Interpupillary Distance) before calibration to ensure a clear and accurate reading from the eye tracking system.

By calibrating the eye tracker every time you use the Vive Focus Vision (or if the headset's fit changes), you'll ensure that the tracking remains precise and responsive throughout your VR sessions.

How Eye Tracking Improves VR Experience:

Eye tracking is a game-changer in virtual reality, contributing significantly to both the realism and performance of VR environments. Here's how it enhances the experience:

1. Enhanced Interactivity:

One of the most immediate benefits of eye tracking is increased interactivity. In traditional VR setups, users must physically reach for objects or rely on controllers to interact with the environment. With eye tracking, the system can detect exactly where you're looking and enable gaze-based interactions. For instance, you can simply look at an object to highlight it, select it, or even initiate a function, which makes the virtual world feel more intuitive and responsive. Eye tracking allows for interactions that mirror natural human behaviors, resulting in smoother and more immersive gameplay.

2. Improved Visual Fidelity:

Eye tracking also plays a key role in foveated rendering, a technique that optimizes the graphical performance of VR experiences. With eye tracking, the system can render only the area you're looking at in full detail, while reducing the resolution of the surrounding areas. This creates the illusion of a high-quality image at a lower computational cost, resulting in smoother performance and less strain on the system. The user's peripheral vision doesn't need to be rendered in high detail, as the eye's natural ability to focus on objects directly in front of it allows for a high-fidelity experience where it matters most.

3. Reduced Motion Sickness:

Motion sickness in VR often occurs when there is a mismatch between what you expect to see and what you actually experience, especially when the graphics are blurry or not rendered in real time. Eye tracking can help alleviate some of this discomfort by ensuring that the visuals always appear clear and aligned with your natural focal points. As the system adjusts to your gaze, the experience feels more natural and less jarring, which can reduce the likelihood of motion sickness during intense VR sessions.

4. More Realistic Virtual Avatars:

In social VR applications, avatars can be made to mirror your eye movements, creating a much more realistic and lifelike representation of yourself in virtual spaces. If you're in a meeting, a game, or an interaction with others in a virtual world, your avatar's gaze will match your own, improving the level of immersion for both you and others. This also opens up new avenues for creating more dynamic, emotionally aware avatars that can display subtle eye movements and reactions to enhance communication and realism.

Setting Up Dynamic Foveated Rendering:

Foveated rendering is one of the most notable features enabled by eye tracking. The basic principle behind foveated rendering is simple: because the human eye has a small region of sharp focus, known as the fovea, it can only see high detail in a small area at any one time. The rest of the visual field is blurred, as the brain doesn't need all the surrounding areas to be as sharp for interpretation.

With eye tracking, the Vive Focus Vision can mimic this biological trait by focusing resources on rendering the area that the user is looking at in full detail. At the same time, it reduces the graphical load in the peripheral areas. Here's how you can set up and optimize dynamic foveated rendering:

1. Activating Foveated Rendering:

Foveated rendering is automatically enabled once eye tracking is calibrated. However, you can adjust the level of foveated rendering through the settings menu. Depending on the headset's performance and the type of content you're using, you can tweak how aggressively the system reduces detail in the peripheral areas.

2. Adjusting the Foveated Rendering Settings:

In the settings, you'll usually find options to control the intensity of foveated rendering. You can opt for mild settings for a less noticeable effect, or go for high intensity to maximize system performance. The higher the intensity, the more resources are freed up for the central region, making the experience smoother, but the trade-off is that the peripheral vision will look more blurred.

3. Impact on Performance:

Foveated rendering greatly improves the headset's performance, particularly for graphically demanding games and applications. By reducing the rendering workload in the peripheral regions, the Vive Focus Vision can maintain higher frame rates, better resolution in the center, and more consistent performance, all without requiring more GPU power.

In conclusion, the combination of eye tracking and foveated rendering significantly enhances the overall VR experience. Eye tracking makes interactions more intuitive, improves visual fidelity, and even reduces motion sickness. Meanwhile, foveated rendering helps optimize performance, making it possible to enjoy a highly detailed VR experience without taxing the system's resources. Together, these technologies set the Vive Focus Vision apart as one of the most advanced headsets available, offering both comfort and efficiency for the modern VR user.

Chapter 7.

Tracking and Movement

One of the key features that enhance the virtual reality experience with the HTC Vive Focus Vision is its advanced tracking capabilities. The system utilizes a combination of hand tracking, controller tracking, and sensor calibration to provide an immersive and seamless experience. This chapter will delve into the setup and calibration of hand tracking, the usage of Vive Focus Vision controllers, tracking modes and sensor calibration, as well as the integration of body tracking with additional accessories to expand the virtual space and interactions.

Hand Tracking Setup and Calibration:

Hand tracking is an essential feature in modern virtual reality (VR), enabling users to interact with virtual environments more naturally. The Vive Focus Vision incorporates hand tracking technology, which allows the system to detect and interpret the movements of the user's hands, even without the use of controllers.

1. Enabling Hand Tracking:

Hand tracking on the Vive Focus Vision is turned on by default in the settings. However, if you wish to enable or disable it, you can access the hand tracking settings in the System Settings menu within the VR interface. Once activated, the system begins tracking the movement of your hands through the use of infrared sensors located on the headset.

2. Calibration Process:

To ensure accurate hand tracking, it is essential to perform a quick calibration. The process starts when you first activate hand tracking and will prompt you to place your hands in a specific position so that the system can detect and calibrate the sensors accordingly. It's important to position your hands within the camera's field of view, making sure that both hands are visible to the sensors. You may need to perform this calibration again if the tracking seems inaccurate or if you change the position of the headset on your head.

3. Adjusting Hand Tracking Sensitivity:

In the settings, users have the ability to adjust the sensitivity of the hand tracking feature. If the tracking feels too sensitive, causing erratic responses, or not sensitive enough, resulting in missed movements, you can fine-tune the tracking sensitivity. This customization helps ensure that hand movements are accurately detected and properly interpreted.

4. Practical Use of Hand Tracking:

Once calibrated, hand tracking allows you to interact with objects and menus by simply moving your hands in the virtual environment. You can point, grab, pinch, and even perform gestures, such as "fist" or "peace" signs, to trigger certain functions or actions. Hand tracking enhances immersion by eliminating the need for physical controllers, creating a more natural and intuitive experience.

Using the Vive Focus Vision Controllers:

While hand tracking is a highly immersive feature, the Vive Focus Vision also comes with controllers for users who prefer more tactile and precise input. The Vive Focus Vision controllers are designed to complement the immersive VR experience with intuitive buttons, triggers, and haptic feedback.

1. Controller Setup:

The Vive Focus Vision includes two controllers, each designed with a set of buttons, a touchpad, and a trigger to interact with the virtual world. When you first use the controllers, you will need to pair them with the headset via Bluetooth. This is typically done through the system settings, where you can go to Device Settings and select Pair Controllers. Once paired, you should see a confirmation on the VR screen.

2. Controller Calibration:

To ensure the controllers provide accurate tracking, they need to be calibrated to the headset's sensor system. Calibration is done automatically when you first connect the controllers. However, if there are tracking issues, recalibration can be done by navigating to the controller settings in the menu and following the on-screen instructions. Calibration ensures the controllers remain aligned with the user's hand movements and are tracked accurately in the virtual space.

3. Using the Controllers in VR:

Once calibrated, you can use the controllers to interact with your environment in various ways. For example, the trackpad allows you to swipe, scroll, or select items, while the trigger buttons can be used for actions like grabbing objects or firing in VR games. The grip buttons can be used for actions like holding objects or initiating specific functions in some apps. The Vive Focus Vision controllers are equipped with haptic feedback, which provides vibration responses based on actions and events, adding an extra layer of immersion when interacting with objects in the virtual space.

4. Tracking the Controllers:

The controllers are tracked using the headset's internal sensors and external cameras. When using the controllers, you should ensure that they remain within the view of the front-facing cameras to maintain optimal tracking. If the controllers move out of view or you experience jittery movement, recalibrating or adjusting the headset position may be necessary to restore full tracking accuracy.

Tracking Modes and Sensor Calibration:

The Vive Focus Vision supports several tracking modes to optimize how the headset and controllers interact with the virtual world. Each mode provides different levels of precision and tracking performance, allowing the system to adapt to various use cases.

1. 6DoF (Six Degrees of Freedom) Tracking:

The Vive Focus Vision supports 6DoF tracking, meaning the system can track movement in three rotational axes (yaw, pitch, and roll) and three translational axes (x, y, and z). This allows the user to move freely in the virtual space and interact with objects in any direction. 6DoF provides a highly accurate and fluid VR experience, especially for action-packed games and simulations.

2. 3DoF (Three Degrees of Freedom) Tracking:

In situations where less complex tracking is required, such as watching videos or interacting with simple environments,

3DoF tracking may be enabled. This mode tracks rotational movements but does not support full movement within the environment. The Vive Focus Vision can switch between these modes automatically based on the activity or user input.

3. Sensor Calibration:

To maintain accurate tracking, the Vive Focus Vision's sensors need to be calibrated regularly. This includes both the internal sensors in the headset and the external tracking cameras. When you first set up the system, the sensors will be automatically calibrated. However, if you notice tracking inconsistencies (e.g., stuttering or drifting), you may need to recalibrate by going to the Sensor Settings section in the system menu and following the recalibration process.

Using Body Tracking with Additional Accessories:

In addition to hand tracking and controller usage, the Vive Focus Vision can be enhanced with additional accessories to incorporate full-body tracking. This allows for more accurate avatar representation and improved interaction in the virtual environment. The system supports a variety of accessories that enhance the body tracking experience.

1. Vive Tracker and Full-Body Tracking:

The Vive Focus Vision supports the use of the Vive Tracker and other compatible accessories, which can be attached to your body or other equipment to track full-body movements. For example, you can use Vive Trackers on your ankles, wrists, or waist to capture more detailed body movements, such as walking, crouching, or gestures that involve the whole body. This level of tracking is particularly useful in VR fitness apps, dance programs, or multiplayer VR games where body movement plays a significant role.

2. Calibration for Full-Body Tracking:

When using body trackers, it's crucial to perform a calibration to ensure that the trackers are accurately mapped to your movements. The calibration process involves positioning the trackers at the designated body points and ensuring the system correctly detects and synchronizes the trackers' positions with the avatar in the virtual world. Once calibrated, your avatar will mimic your real-world movements, making the virtual experience feel even more lifelike.

3. Body Tracking Settings:

Once you have connected additional trackers, you can adjust the body tracking settings within the system menu. This includes calibrating the trackers, adjusting the level of precision, and tweaking the way the trackers interact with your avatar. These settings can be fine-tuned to suit specific activities or applications.

In conclusion, the Vive Focus Vision offers a highly advanced tracking system that can handle hand tracking, controller tracking, 6DoF, and even full-body tracking with additional accessories. By properly setting up and calibrating the tracking systems, you can ensure a fluid and accurate VR experience that allows you to move and interact naturally in the virtual space. The flexibility and adaptability of the tracking modes make the Vive Focus Vision a versatile tool for both casual users and professional applications, offering a truly immersive VR experience.

Chapter 8.

Maintenance & Care

Maintaining the HTC Vive Focus Vision headset is essential for ensuring optimal performance, longevity, and a consistently enjoyable VR experience. Like any piece of advanced technology, proper care and regular maintenance help protect your investment and ensure the system functions smoothly over time. This chapter will cover how to clean and store the headset, protect the lenses and display, perform firmware and software updates, troubleshoot common issues, and care for the battery and charging process.

Cleaning and Storing the Headset:

Cleaning your HTC Vive Focus Vision regularly ensures that it remains in top condition and helps to avoid dirt, dust, or oils from damaging the device. Proper cleaning and storage also contribute to its comfort and long-lasting performance.

1. Cleaning the Headset:

Before cleaning, always power off your headset to prevent any accidental activation or damage to internal components. Use a soft, lint-free cloth to wipe down the exterior of the headset, including the frame and sensors. For stubborn spots, slightly dampen the cloth with water or a mild cleaning solution designed for electronics. Avoid using harsh chemicals, alcohol, or abrasive materials, as they may damage the headset's surface or finish.

For the face cushion, it's essential to clean it frequently since it comes into direct contact with your skin. Some headsets, including the Vive Focus Vision, come with removable face cushions, which can be gently wiped down or washed (if machine washable). For non-removable cushions, use a disinfectant wipe or a cloth with a diluted cleaning solution to keep the area fresh and hygienic.

2. Storing the Headset:

When not in use, store the headset in a cool, dry place. Avoid leaving it in direct sunlight or in areas with extreme temperatures, as this can damage the lenses or affect the headset's performance. If you're not using it for an extended period, it's a good idea to store the headset in a protective case or bag to shield it from dust and accidental damage. A protective case can also prevent scratches and physical wear from everyday handling.

Protecting the Lenses and Display:

The lenses and display are among the most delicate parts of your HTC Vive Focus Vision headset, and proper care is essential to avoid scratches, smudges, or any permanent damage. Here are the key steps to protect the lenses and display:

1. Lens Cleaning:

Lenses are highly sensitive and must be treated gently. To clean the lenses, use a microfiber cloth. Lightly wipe in a circular motion to remove fingerprints, smudges, or dust. Never apply cleaning solution directly onto the lenses. Instead, spray a small amount of cleaner onto the cloth to avoid drips and potential damage to the lens surface.

If your headset has lens protectors (usually sold separately), use them to shield the lenses when the headset is not in use. These can be particularly useful if you frequently transport your headset.

2. Avoid Scratches:

Always store the headset in a case or on a protective stand that keeps it from coming into contact with hard surfaces. If storing it on a table or shelf, be mindful of other objects that could cause scratches. For example, placing the headset face-up when not in use can prevent accidental contact with sharp or rough surfaces.

3. Display Protection:

The LCD screens of the Vive Focus Vision should be kept free from dust, dirt, and scratches. The screens are vulnerable to physical damage if not handled carefully. Avoid using sharp objects near the display, and never press directly on the screen with your fingers or any tool. Consider applying a screen protector specifically designed for VR headsets if you're concerned about potential damage.

Firmware and Software Updates:

To ensure your HTC Vive Focus Vision operates at peak performance, it's important to keep both the firmware and software up to date. Regular updates can improve functionality, add new features, fix bugs, and enhance compatibility.

1. Checking for Firmware Updates:

The Vive Focus Vision uses OTA (over-the-air) updates to update firmware. These updates typically include improvements for system performance, security patches, and new features. You'll be notified when an update is available, and you can follow on-screen instructions to install it. Make sure your headset is fully charged before performing any update to avoid interruptions during the process.

2. Software Updates:

In addition to firmware updates, the Vive Focus Vision apps and associated software (like the Vive Focus app) will also require occasional updates. These updates may include bug fixes, new VR experiences, or enhancements to existing features. You can download these updates through the Vive Focus app or through the Viveport platform on the headset. Ensure that the Wi-Fi connection is stable during the update process to prevent any issues.

3. Performing a Manual Update:

If your headset doesn't automatically prompt you to update or if you're experiencing difficulties with automatic updates, you can manually check for updates in the Settings menu under System and select Software Update. Here, you can also review the update history and schedule future updates if necessary.

Troubleshooting Common Issues:

Over time, you may encounter a few issues with your HTC Vive Focus Vision. Many common problems can be resolved with a simple troubleshooting process.

1. No Display or Blank Screen:

If the screen appears blank or black, ensure that the headset is powered on and that the connections to your device are secure. If the display still doesn't appear, try restarting the headset. A reset to factory settings may be required if the issue persists.

2. Poor Tracking or Controller Issues:

If the controllers aren't tracking accurately or hand tracking seems off, recalibrate the headset sensors. Make sure that the area you're using the headset in is free from obstructions and that there is sufficient lighting. Poor lighting or reflective surfaces can interfere with tracking.

3. Audio Problems:

If you're experiencing audio issues (e.g., no sound, low volume, or distorted sound), check that the audio output is correctly set and the volume is turned up on both the headset and any connected devices. Ensure that no debris or dirt is blocking the headset's audio speakers.

4. Overheating:

Overheating can occur if the headset is used for extended periods in a warm environment. To resolve this, turn off the headset and let it cool down. If the issue persists, check for any blocked ventilation or consider reducing usage duration to prevent overheating.

Battery Care and Charging Tips:

The Vive Focus Vision uses a rechargeable battery, and proper charging practices are essential for maintaining its battery health and ensuring optimal performance.

1. Charging the Headset:

To charge your headset, use the USB-C charging cable included with the device. Plug one end into the headset and the other into a power source (such as a wall charger or power bank). Make sure to use a charger with the correct output specifications to avoid overcharging or damaging the battery.

2. Charging Tips:

Avoid charging the headset overnight or leaving it plugged in for extended periods. This can prevent the battery from overcharging and help preserve its longevity.

If you plan not to use the headset for a while, store it with the battery at a 50% charge to prevent battery degradation.

Try to keep the battery between 20% to 80% charged for optimal battery health. Charging it fully to 100% or draining it completely to 0% frequently can shorten battery life over time.

3. Battery Maintenance:

Over time, the battery will naturally degrade and hold less charge. If you notice that the headset's battery drains unusually fast, it may be time to replace it. HTC typically provides guidelines for replacing the battery, and this should be done by a certified technician to avoid any damage.

Conclusion:

By following these maintenance and care tips, you can keep your HTC Vive Focus Vision headset in optimal condition. Regular cleaning, proper storage, and caring for the lenses and display will ensure the headset remains free of damage. Staying on top of firmware and software updates keeps your device performing at its best, and understanding how to troubleshoot common issues will help you resolve any potential problems. Lastly, practicing proper battery care and charging techniques will help extend the life of your device. Taking good care of your Vive Focus Vision will enhance your overall VR experience and protect your investment for years to come.

Chapter 9.

Advanced Features

The HTC Vive Focus Vision is equipped with a suite of advanced features that enhance the user experience, making it one of the most versatile and high-performance mixed-reality headsets on the market. These features allow you to tailor the headset to your personal preferences, expand its functionality, and integrate additional hardware. In this chapter, we'll explore some of the more advanced features available on the Vive Focus Vision, including customizing the user interface, setting up multiple user profiles, utilizing face and body tracking with add-ons, and taking advantage of the Vive Wired Streaming Kit.

Customizing the User Interface:

The Vive Focus Vision offers a flexible and user-friendly interface that can be customized to suit individual needs. Customizing the user interface (UI) helps streamline your VR experience, making it easier to navigate, access content, and optimize the layout based on how you use the headset.

1. Adjusting the UI Layout:

The Vive Focus Vision UI features a range of layouts and shortcuts that can be personalized. For example, you can choose to display the home screen apps in a grid or list format, depending on your preferences. You can also prioritize apps and tools, allowing you to quickly access your most-used applications or features. To customize the layout, simply go to the settings menu and select Display or User Interface. Here, you can drag and drop shortcuts and adjust the UI elements to better suit your workflow.

2. Changing Themes and Colors:

For users who prefer a more personalized aesthetic, the Vive Focus Vision allows you to adjust the theme and color scheme of the interface. Whether you prefer a dark mode for reducing eye strain or a light mode for better visibility, you can select your preferred option in the Settings > Display menu. Additionally, certain third-party apps may allow you to further personalize the UI within those specific applications, giving you full control over how your experience looks and feels.

3. Accessing Shortcuts and Gesture Controls:

The Vive Focus Vision also supports gesture-based controls, allowing you to perform various actions with hand movements. These gestures can be customized to perform different functions, such as opening specific apps, controlling media playback, or adjusting settings. You can configure these gestures through the Settings > Accessibility menu. This customization lets you interact with the VR environment in a more intuitive and efficient way, reducing reliance on physical controllers.

Multi-User Setup and Profiles:

The Vive Focus Vision is designed to support multiple users, making it an excellent option for families, businesses, or shared environments. The multi-user setup allows each person to create their own personalized profile, preserving individual settings, preferences, and VR experiences.

1. Creating User Profiles:

To set up a new user profile, go to the Settings > Users menu. From here, you can create and manage different profiles by selecting Add New User. Each user will have their own customized settings, such as UI layout, preferred apps, and saved content. This feature ensures that multiple users can seamlessly switch between profiles without affecting the other's settings or experiences.

2. Switching Between Profiles:

Switching between user profiles is straightforward. Simply navigate to the User Profile section in the settings and select the profile you want to use. The headset will automatically adjust to the selected user's preferences. This feature is particularly useful in environments where multiple people use the same device, ensuring that each user's VR experience remains personalized.

3. Parental Controls and Content Restrictions:

For families or educational environments, the Vive Focus Vision allows administrators to set up parental controls to restrict access to certain content or features. You can limit access to age-appropriate applications, websites, or games, ensuring that the headset is safe for younger users. These restrictions can be easily managed through the Parental Controls section in the Settings.

Using Face and Body Tracking (with Add-ons):

One of the standout features of the Vive Focus Vision is its ability to support additional tracking devices, such as the Vive Facial Tracker and the Vive Body Tracker. These add-ons significantly enhance the immersion of your VR experience by enabling facial expressions and full-body movement tracking.

1. Vive Facial Tracker:

The Vive Facial Tracker is an add-on accessory that attaches to the front of the headset, enabling facial tracking that captures your expressions in real-time. This feature allows for more immersive social interactions in VR, as your virtual avatar can mirror your facial movements, such as smiling, frowning, or raising an eyebrow. It's especially beneficial for virtual meetings, social VR apps, and games that require emotional expressions.

To use the Vive Facial Tracker, simply attach it to the designated mounting point on the headset and follow the calibration steps in the Settings > Facial Tracking menu.

Calibration typically involves positioning your face properly to ensure that the tracker captures your expressions accurately. Once set up, your avatar in supported applications will reflect your real-time facial movements.

2. Vive Body Tracker:

The Vive Body Tracker is an additional accessory that tracks full-body movements, including the motion of your arms, legs, and torso. This tracking enhances the overall realism of VR by capturing a wide range of movements, allowing your virtual avatar to move as you do in real life. The Vive Body Tracker is particularly useful in fitness applications, motion-sensing games, and any VR experience that requires full-body interaction.

Setting up the Vive Body Tracker involves placing the tracking sensors on your body typically on your waist, feet, and wrists using the provided straps. Once connected to the Vive Focus Vision, the sensors will track your movements and relay them to the VR environment. Calibration is important to ensure accurate tracking, and this can be done through the Settings > Body Tracking menu.

3. Synchronizing with VR Apps:

After setting up the facial and body trackers, you'll need to sync them with compatible VR applications. Many VR games and experiences already support advanced tracking features, including those for facial and full-body movements. Once synchronized, your avatar in supported apps will accurately replicate your real-life movements, providing an incredibly immersive experience.

Using the Vive Wired Streaming Kit:

While the Vive Focus Vision is primarily a standalone VR headset, it also supports PC VR functionality through the use of the Vive Wired Streaming Kit. This kit enables you to stream PC VR content directly to your headset using a DisplayPort connection, providing access to more complex VR applications and games.

1. Setting Up the Vive Wired Streaming Kit:

The Vive Wired Streaming Kit consists of a DisplayPort cable and a converter that connects the Vive Focus Vision to your PC. To set up the kit, first, connect the DisplayPort cable to your PC's graphics card, and then connect the other end to the provided converter. Next, plug the converter into the headset's USB-C port. This setup allows the Vive Focus Vision to access VR content from SteamVR and Viveport, expanding the library of experiences you can enjoy.

2. Configuring PC VR on the Vive Focus Vision:

Once the Wired Streaming Kit is connected, the Vive Focus Vision will automatically recognize the connection. You'll need to configure the PC VR settings within the SteamVR and Viveport apps to ensure that the content is compatible with the Vive Focus Vision. This involves adjusting the display settings, resolution, and other graphics-related parameters to optimize the experience for your headset. Make sure to check that your PC meets the required specifications for VR streaming.

3. Switching Between Standalone and PC VR Mode:

With the Vive Wired Streaming Kit, you can easily switch between the Vive Focus Vision's standalone mode and PC VR mode. To switch, simply disconnect the wired streaming kit when you're done with PC VR content, and the headset will revert to its standalone mode. This flexibility gives you the best of both worlds access to high-quality PC VR experiences and the convenience of a portable, standalone VR system.

Conclusion:

The advanced features of the HTC Vive Focus Vision allow you to fully customize your VR experience, whether through the user interface, multi-user profiles, or additional accessories like face and body tracking. The ability to personalize settings, switch between user profiles, and integrate advanced tracking features elevates the immersion and functionality of the device. Additionally, the Vive Wired Streaming Kit expands the headset's potential by enabling you to access high-quality PC VR content, making it a versatile option for both standalone and PC-based VR experiences. By taking full advantage of these advanced features, you can optimize your HTC Vive Focus Vision to meet your unique needs and preferences, enhancing your VR journey to new levels of immersion and customization.

Chapter 10.

VR Gaming and Applications

The HTC Vive Focus Vision offers a wide array of immersive virtual reality (VR) experiences, from high-octane gaming to social interactions in virtual worlds. Whether you're a gamer or simply looking for a fun way to explore new environments, this headset supports a vast library of content through platforms like Viveport and SteamVR. In this chapter, we'll guide you through installing and playing VR games, using Viveport and SteamVR, exploring recommended apps and games, and delving into the world of social VR apps like VRChat.

Installing and Playing VR Games:

Getting started with VR gaming on the Vive Focus Vision is straightforward. The headset supports a variety of content, whether you prefer standalone VR titles or PC VR games streamed through your computer. Here's how to install and play VR games on the Vive Focus Vision:

1. Standalone VR Games:

The Vive Focus Vision is capable of running standalone VR games directly from the headset without needing a connected PC. To install these games, you'll need to use the Viveport store. Begin by opening the Viveport app from the headset's main interface. You can browse through different categories such as action, adventure, simulation, and puzzle games. Each game will have its own page with information on the gameplay, controls, and compatibility.

To install a game, simply select the game you're interested in, click the Download or Purchase button, and follow the on-screen instructions. Once the download is complete, the game will appear in your library, ready for play. Launch the game directly from the Viveport interface, and enjoy a seamless VR experience.

2. PC VR Games:

If you wish to access more graphically intensive VR games or have a library of PC VR titles, you can connect the Vive Focus Vision to a PC using the Vive Wired Streaming Kit or wirelessly using SteamVR's built-in wireless streaming options. Once the connection is established, you can browse your Steam or Viveport library and launch the games as you would on a traditional VR headset.

Before playing PC VR games, ensure your computer meets the necessary specifications and that the Vive Focus Vision is connected correctly via either wired or wireless streaming. The headset will automatically recognize PC VR mode, and you'll have access to all of your PC-based content.

How to Use Viveport and SteamVR:

The Vive Focus Vision can connect to both Viveport and SteamVR, providing access to an extensive range of VR games, apps, and experiences. Here's how to use each platform effectively:

1. Viveport:

Viveport is HTC's proprietary platform for VR content, offering a subscription service called Viveport Infinity, which grants access to a vast catalog of VR games and apps for a monthly fee. To use Viveport, follow these steps:

Download the Viveport App:

If it's not already installed, head to the Viveport store within the headset's menu and download the app. It will serve as your gateway to VR content.

Browse and Install Content:

Once installed, open the Viveport app and sign in with your HTC account. Browse through the extensive library of VR content, including games, educational apps, creative tools, and more. You can filter content by categories such as "Best Sellers," "New Releases," or "Recommended for You."

Viveport Infinity:

If you're subscribed to Viveport Infinity, you'll have unlimited access to thousands of VR experiences. You can download as many apps and games as you want without additional charges, making it an excellent option for avid VR users.

To purchase or download a title, click on the game or app's page, and follow the on-screen instructions to complete the process. Games will be installed directly on the headset, and you can find them in your library.

2. SteamVR:

SteamVR is one of the most popular VR platforms, hosting a large catalog of games, apps, and experiences. The Vive Focus Vision supports SteamVR games when connected to a PC via wired or wireless streaming. Here's how to get started:

Install SteamVR:

If you haven't already, download and install SteamVR on your PC. This can be done through the Steam client, which is available for free. SteamVR will allow your PC to interface with VR headsets, including the Vive Focus Vision.

Connecting Vive Focus Vision to PC:

For wired connections, use the Vive Wired Streaming Kit. For wireless streaming, ensure your PC and Vive Focus Vision are on the same Wi-Fi network. SteamVR will detect the headset automatically once it's properly connected.

Browse SteamVR Content:

With SteamVR running, browse the Steam store to find compatible VR games. When you purchase a game, it will appear in your library, ready for installation. After installation, simply launch the game, and it will start in VR mode, displaying content on the Vive Focus Vision headset.

SteamVR also supports numerous interactive tools and environments, such as mixed-reality experiences, fitness apps, and social VR, expanding the ways you can engage with virtual environments.

Recommended Apps and Games:

There's no shortage of incredible VR games and applications to explore on the Vive Focus Vision. Whether you're into fast-paced action games or immersive exploration, you'll find something that fits your taste. Here are some recommendations:

1. VR Games:

Beat Saber:

One of the most popular rhythm-based games in VR, Beat Saber has players slice through blocks to the beat of the music. It's available on both Viveport and SteamVR, and it's an excellent way to get into VR fitness.

Superhot VR:

A first-person shooter with a unique twist time only moves when you move. It's a fun and strategic experience, perfect for those looking for action-packed gameplay.

The Walking Dead:

Saints & Sinners: For fans of horror and action, this game lets you step into a post-apocalyptic world filled with zombies. It's an intense and immersive survival experience.

2. VR Apps:

Tilt Brush:

An art app that lets you paint in three-dimensional space. If you're creative, Tilt Brush will let you bring your artwork to life in the virtual world.

Wander:

A virtual travel app that lets you explore real-world locations using Google Street View. It's perfect for armchair travelers or anyone looking to explore the world in a new way.

Viveport VR:

Not only a store for purchasing VR content, but Viveport VR also offers curated experiences, allowing you to explore a wide range of environments from science fiction to fantasy and everything in between.

3. Fitness and Wellness Apps:

FitXR:

A fitness-focused VR game that offers cardio and boxing workouts. It's a fun way to stay in shape while immersed in a virtual environment.

VZFit:

A cycling app that allows you to cycle through virtual landscapes. It integrates with fitness equipment and offers an enjoyable way to exercise from home.

Playing VRChat and Other Social Apps:

One of the most exciting aspects of the Vive Focus Vision is the ability to connect with others in virtual worlds. VRChat is one of the most popular social VR apps, and it's available on both Viveport and SteamVR. Here's how you can get started:

1. Installing VRChat:

Download VRChat from the Viveport or SteamVR store, depending on whether you're using the standalone mode or PC VR streaming. VRChat is free to play and allows you to explore a massive online world, interact with others, and even create your own avatars and environments.

2. Creating an Avatar:

In VRChat, you can create a custom avatar, which can be made from scratch or downloaded from the community. Customize your avatar's look, from the color of its skin to its clothing and accessories. This adds a personal touch to your VR social experience, as you can express yourself however you choose.

3. Social Interaction:

Once in VRChat, you can meet people from around the world. VRChat offers a variety of worlds to explore, from relaxing spaces to action-packed games. You can join other players in shared spaces, chat, play games, or even attend virtual events.

4. Other Social Apps:

Apart from VRChat, several other social VR platforms exist, such as AltspaceVR and Rec Room. These apps allow users to meet new people, play multiplayer games, and attend virtual meetups or live events. All of these apps support multiplayer interactions, ensuring that your social VR experience remains lively and engaging.

Conclusion:

The HTC Vive Focus Vision provides endless opportunities for gaming, socializing, and discovering new VR worlds. By exploring Viveport, SteamVR, and social apps like VRChat, you can immerse yourself in a vast library of content. Whether you're playing solo or with others, the Vive Focus Vision brings the virtual world to life, making it a must-have tool for both entertainment and social interaction in VR. With its versatility, compatibility, and support for a wide range of content, the Vive Focus Vision is the ideal VR device for any enthusiast.

Chapter 11.

Connecting with Other Devices

The HTC Vive Focus Vision is designed to offer seamless integration with a variety of other devices to enhance your VR experience. Whether you're looking to use wireless audio, pair additional accessories, or cast your virtual reality content to a larger screen, the headset provides a multitude of connectivity options. In this chapter, we'll walk you through pairing Bluetooth devices like headphones and controllers, using the Vive Focus Vision with other VR accessories, and casting VR content to other screens.

Pairing with Bluetooth Devices (Headphones, Controllers):

One of the most useful features of the HTC Vive Focus Vision is its ability to pair with Bluetooth devices such as wireless headphones and controllers. This ensures that you can experience high-quality audio and control your VR environment without the limitations of wires.

Pairing Bluetooth Headphones:

Listening to immersive sound is a vital part of any virtual reality experience. The Vive Focus Vision supports Bluetooth pairing, so you can easily connect wireless headphones to the headset for a cable-free experience. Here's how to pair your Bluetooth headphones:

1. Turn on Your Bluetooth Headphones:

Ensure that your Bluetooth headphones are powered on and in pairing mode. If you're unsure how to do this, consult your headphones' manual, as the process can vary depending on the model.

2. Access Bluetooth Settings:

Put on the Vive Focus Vision headset and navigate to the Settings menu from the main screen. Look for the Bluetooth option within the connectivity settings.

3. Enable Bluetooth Pairing:

In the Bluetooth settings menu, enable the option to start searching for nearby devices. The Vive Focus Vision will search for any Bluetooth-enabled devices in range, including your headphones.

4. Select Your Headphones:

Once the headset detects your headphones, select them from the list of available devices. A confirmation message will appear on the screen once the pairing is successful.

5. Test the Audio:

After pairing, check to see if the audio is working correctly by playing some VR content or navigating through the system menu. If needed, adjust the volume through the Sound Settings in the system menu.

This process ensures a smooth audio experience without the inconvenience of wires, allowing for greater freedom of movement while in VR.

Pairing Bluetooth Controllers:

The Vive Focus Vision also supports Bluetooth controllers, offering an even more immersive and interactive experience. You can pair your headset with the Vive Focus controllers or any compatible third-party controllers for enhanced navigation and gameplay. To pair your Bluetooth controllers:

1. Prepare the Controllers:

Make sure that the controllers are charged and powered on. If using Vive Focus controllers, ensure they are in pairing mode by holding the power button for a few seconds until the LED indicator blinks.

2. Navigate to the Controller Settings:

In the Settings menu of the Vive Focus Vision, select the Controllers option, which will allow you to view and manage paired controllers.

3. Pair the Controllers:

After enabling Bluetooth, the headset will automatically search for nearby Bluetooth controllers. Select your controller from the list to initiate the pairing process.

4. Confirm Pairing:

A confirmation message will appear once the controllers are paired successfully. You can test the controllers by navigating through the menus or launching a VR app. Ensure that the controllers respond correctly to inputs like pressing buttons, pointing, and tracking movements.

With controllers paired, you can navigate menus, interact with VR applications, and immerse yourself in games with ease.

Using the Vive Focus Vision with Other VR Accessories:

While the Vive Focus Vision is a standalone headset, it also supports a wide range of additional VR accessories that can enhance your experience. From facial trackers to ultimate trackers for body movement, you can integrate these accessories to create a more immersive and customized VR experience. Below are the main accessories compatible with the Vive Focus Vision:

1. Facial Tracker

The Facial Tracker is an optional accessory that tracks the movements of your face and allows for more personalized avatars or expressions in VR. It uses infrared sensors to track facial muscles and translate those movements into the virtual world, adding an extra layer of realism and immersion.

Setting Up the Facial Tracker:

To use the Facial Tracker with the Vive Focus Vision:

First, make sure the facial tracker is fully charged.

Connect the facial tracker to the headset using the USB-C port or Bluetooth, depending on the model.

In the Settings menu, select the Facial Tracker option to enable its functionality.

Follow the on-screen instructions to calibrate the tracker, ensuring it's properly positioned for accurate tracking.

Once calibrated, you'll be able to see real-time expressions, including mouth movements and eye tracking, applied to your VR avatar.

2. Ultimate Trackers:

The Ultimate Trackers are a set of motion-tracking devices that can be used to track your body movements and integrate them into the virtual world. These trackers are useful for applications like fitness VR games, motion capture for virtual production, and other immersive experiences where full-body tracking is required.

Setting Up the Ultimate Trackers:

Attach the ultimate trackers to your body using straps (usually worn on your wrists, ankles, or waist).

Ensure each tracker is charged and powered on.

Connect the trackers to the Vive Focus Vision by pairing them via Bluetooth.

Use the Tracker Calibration option in the Settings menu to calibrate the devices and ensure proper motion tracking.

Once configured, the ultimate trackers allow the Vive Focus Vision to capture your movements in full 3D space, translating them into the virtual world for a truly immersive experience.

Casting VR Content to Other Screens:

Another useful feature of the HTC Vive Focus Vision is the ability to cast your VR content to other screens, such as a TV, monitor, or even a mobile device. This feature is perfect for sharing your VR experiences with others in the room or displaying your VR gameplay for a broader audience.

1. Casting to a Smart TV

Casting VR content to a smart TV is a simple process and works seamlessly with compatible models. Here's how to cast your VR content:

1. Ensure Compatibility:

Make sure your smart TV supports Google Chromecast or has built-in casting functionality (many smart TVs do).

2. Connect Both Devices to the Same Network:

Ensure both your Vive Focus Vision headset and the smart TV are connected to the same Wi-Fi network.

3. Enable Casting:

In the Settings menu of your Vive Focus Vision, look for the Cast option. Tap this option to enable casting mode.

4. Select Your TV:

The system will automatically detect nearby casting devices, including your smart TV. Select your TV from the list.

5. Start Casting:

Once connected, the content from your Vive Focus Vision will be mirrored onto the TV screen. You can now enjoy sharing your VR experiences with others in real-time.

2. Casting to a Mobile Device or PC

If you want to cast your VR experience to a mobile device or PC, the process is quite similar. Here's how to do it:

1. Use Vive Sync App:

Install the Vive Sync app on your mobile device or PC. This app supports casting from your Vive Focus Vision to other devices.

2. Connect Devices to the Same Network:

Ensure both the Vive Focus Vision and the mobile device or PC are connected to the same Wi-Fi network.

3. Start Casting:

Open the Vive Sync app and select the "Cast" option. Then, select the target device (either a mobile or PC) to begin streaming the content from your Vive Focus Vision.

This feature is particularly useful for content creators who want to showcase their VR gameplay or simply for social VR experiences where others can watch along.

Conclusion:

The HTC Vive Focus Vision offers extensive connectivity options that greatly enhance the flexibility and functionality of the headset. By pairing Bluetooth devices like headphones and controllers, integrating accessories like facial and ultimate trackers, and casting VR content to other screens, users can create a truly immersive and social VR experience. These features provide flexibility for various use cases, from personal gaming and entertainment to content creation and social VR interactions. By leveraging these connection capabilities, you can elevate your virtual reality experience to new heights.

Chapter 12.

Troubleshooting and FAQs

As with any advanced piece of technology, issues may arise while setting up or using the HTC Vive Focus Vision. Fortunately, most problems can be resolved with some simple troubleshooting steps. This chapter aims to guide you through common setup issues, display and performance problems, and issues related to eye tracking, connectivity with PC VR, and DisplayPort functionality. Whether you're facing difficulties during initial setup or while using the headset, this guide will provide practical solutions.

Common Setup Issues and Solutions:

Setting up the HTC Vive Focus Vision should be a straightforward process, but in some cases, users may encounter common issues. These problems typically relate to initial configurations, software installation, or hardware connections. Below are the most frequently reported setup problems and their solutions:

1. Headset Not Powering On

Solution:

If the headset doesn't power on, first ensure that it is fully charged. Use the provided charging cable and adapter to plug the headset into a power source. If the headset still doesn't power on, perform a soft reset by holding down the power button for about 10 seconds, then try powering it on again. If the issue persists, check for damage to the power button or charging port.

2. Vive Focus Vision Not Detecting Wi-Fi Network

Solution:

If your headset isn't detecting your Wi-Fi network during setup, ensure that you are within range of your router and that the Wi-Fi network is functioning correctly. Try restarting your router or resetting the Wi-Fi connection on the Vive Focus

Vision by going to Settings > Wi-Fi and reconnecting. If you're on a 5GHz network and the headset is having trouble, try connecting to a 2.4GHz network instead.

3. App Installation Issues

Solution: If you have trouble downloading or installing the Vive Focus Vision app or any other software, check your internet connection and ensure the device has enough storage. If the app is still not installing, try restarting the headset and reinstalling the app. In some cases, resetting the headset to factory settings may resolve persistent software issues.

Resolving Display and Performance Problems:

The HTC Vive Focus Vision offers stunning visuals with its high-resolution displays. However, there are times when users may experience display or performance-related issues. Below are the common problems and how to address them.

1. Screen Flickering or Blackouts

Solution:

Flickering or blackouts on the screen can occur for several reasons, including loose cables, software bugs, or hardware issues. First, ensure the headset is correctly connected and that all cables are secure. If you're using it in PC VR mode with a wired connection, check the DisplayPort or HDMI cables. If you're using wireless, try reducing the distance between the headset and the router to improve signal strength. Restarting the headset can also clear up minor glitches that may cause flickering.

2. Low Frame Rates and Lag

Solution:

Lag or low frame rates can result in a suboptimal VR experience. If you're using the Vive Focus Vision in standalone mode, try closing background apps and reducing the quality of visuals by adjusting the settings within the app you're using. In PC VR mode, low frame rates can occur due to insufficient hardware performance on the connected PC.

Ensure that your PC meets the minimum requirements for VR gameplay, and make sure that your PC's GPU drivers are up to date. Reducing the resolution or lowering in-game graphical settings can help improve performance.

3. Blurry or Distorted Display

Solution:

If the display appears blurry or distorted, check the lens for smudges or dust, as these can affect image clarity. Clean the lenses gently with a microfiber cloth. If the image is still unclear, adjust the IPD (Interpupillary Distance) setting to match the distance between your eyes. To adjust IPD, go to the settings and modify the lens position until the display appears crisp. Also, ensure the headset is positioned correctly on your head for optimal focus.

Eye Tracking Not Working Properly:

Eye tracking is a key feature of the HTC Vive Focus Vision, but there may be occasions when it does not function as expected. Here are common issues with eye tracking and how to fix them.

1. Calibration Issues

Solution:

Eye tracking requires proper calibration to function correctly. If the eye tracking isn't responding or is inaccurate, try recalibrating it by following the on-screen instructions. To do so, go to the Settings > Eye Tracking and select Calibrate. Ensure you are sitting in a well-lit area, and position the headset correctly on your head to facilitate accurate tracking.

2. Eye Tracking Not Responding or Inaccurate

Solution:

If the eye tracking is not responding or appears inaccurate, ensure that the headset is positioned properly and that there are no obstructions between the sensor and your eyes. The sensors may not work well if you are wearing glasses or if there is a glare on the lenses. In such cases, try removing your glasses or adjusting the lighting in the room to reduce glare. Additionally, check if the headset's firmware is up to date, as software updates often address issues with eye tracking.

Connectivity Problems with PC VR:

The HTC Vive Focus Vision can be connected to a PC to access PC VR content, but users may encounter connectivity issues when trying to stream content. The most common problems involve the DisplayPort connection, SteamVR, or Viveport compatibility. Here's how to resolve these issues.

1. Vive Focus Vision Not Connecting to PC

Solution:

If the Vive Focus Vision is not connecting to your PC, first ensure that your PC meets the necessary requirements for VR gameplay. Verify that the PC's GPU and USB ports are functioning and compatible with the Vive Focus Vision. Try using a different USB cable or port to connect the headset to the PC. Ensure the Vive Focus Vision drivers are installed on your PC, and restart both the headset and the computer. Additionally, check if the headset is properly detected in SteamVR or Viveport, and make sure that the latest software updates are installed.

2. SteamVR or Viveport Not Detecting the Headset

Solution:

If SteamVR or Viveport is not detecting the Vive Focus Vision, make sure the headset is connected properly and is powered on. If using a wireless connection, ensure that the Vive Focus Vision is on the same Wi-Fi network as the PC. If using a wired connection, check the DisplayPort or USB connection for any issues. Restart both the Vive Focus Vision headset and your PC to reset the connection. If the issue persists, try reinstalling SteamVR or Viveport and update to the latest version.

DisplayPort Issues:

If you're using the Vive Focus Vision in PC VR mode and experiencing issues with the DisplayPort connection, here are some solutions to consider.

1. No Signal on DisplayPort

Solution:

If the headset displays a "no signal" error or a black screen when connected via DisplayPort, first verify that the DisplayPort cable is securely connected to both the PC and the headset. Ensure that your PC's GPU supports DisplayPort 1.4 or higher. If using an adapter, make sure it's compatible with your system. Sometimes, DisplayPort issues can be resolved by restarting the PC and headset.

2. DisplayPort Performance Issues

Solution:

If the display appears choppy or laggy during gameplay, try reducing the resolution and refresh rate in your PC VR settings. Lowering these settings can help maintain stable performance. Additionally, check for any GPU driver updates, as outdated drivers can cause performance degradation. If using an external adapter, ensure it is functioning correctly and supports the necessary performance requirements for VR gaming.

Conclusion:

The HTC Vive Focus Vision is an advanced piece of technology that can occasionally experience setup, display, and connectivity issues. However, most of these problems can be easily resolved with some basic troubleshooting steps. Whether you are dealing with power issues, eye tracking inaccuracies, or connectivity problems with your PC, the solutions provided in this chapter should guide you toward resolving the problem quickly and getting back to your immersive VR experience. For persistent issues, always ensure that your device is updated to the latest firmware and drivers, and consult HTC's customer support for further assistance.

Chapter 13.

Safety & Health

Virtual Reality (VR) provides an incredible experience, immersing users in virtual environments that engage the senses in ways that traditional media cannot. However, prolonged use of VR headsets like the HTC Vive Focus Vision can lead to health issues such as eye strain, motion sickness, and physical discomfort. This chapter will focus on the best practices for using VR safely and comfortably, offering tips to reduce potential health risks while ensuring an enjoyable and productive experience.

VR Safety Tips for Comfortable Use:

When using VR, it's essential to prioritize safety to avoid physical and mental discomfort. Here are several general VR safety tips to ensure a comfortable experience.

1. Adjust the Headset Properly

One of the most crucial steps for comfortable VR use is ensuring the headset fits properly. A poorly adjusted headset can lead to eye strain, neck discomfort, and blurry visuals. When setting up your HTC Vive Focus Vision, make sure to adjust the straps to fit snugly on your head without being too tight. The center of the display should align with your eyes to ensure that you get the best view and prevent unnecessary strain. Adjust the Interpupillary Distance (IPD) to match your eye distance for sharp, clear visuals.

2. Start Slowly

If you are new to VR, or if it's been a while since your last session, it's important to start slowly and gradually increase your time in VR. Begin with shorter sessions of 15 to 20 minutes, particularly in the early stages, to allow your body to acclimate to the virtual world. This approach will also reduce the likelihood of experiencing motion sickness or discomfort from overstimulation.

3. Clear Your Play Area

A safe environment is critical for VR gaming or other experiences. Before starting, ensure that your play area is free of obstacles, furniture, and anything that might cause you to trip or bump into. Many VR systems, including the Vive Focus Vision, offer room setup guides that help define a play space and alert you when you're getting too close to the boundaries. Follow these guidelines to avoid accidents and ensure you can move around freely without hitting objects.

4. Use Safety Zones and Boundaries

Most VR headsets, including the Vive Focus Vision, feature a safety zone that warns you when you're near the edges of your play area. Be sure to enable and adjust these boundary alerts before using your headset. This feature helps you stay aware of your surroundings and can prevent accidental collisions.

Reducing Eye Strain and Motion Sickness:

While VR is incredibly immersive, extended use can sometimes lead to eye strain or motion sickness. These issues can diminish the enjoyment of your VR experience, but with a few simple adjustments, you can reduce their impact.

1. Take Regular Breaks

To minimize eye strain and reduce the risk of motion sickness, take breaks regularly. A good rule of thumb is to take a 10-15 minute break every 30-45 minutes of VR use. During this break, look at objects in the real world to give your eyes a chance to relax and focus on something different. If you feel eye fatigue setting in, pause and rest your eyes before continuing.

Additionally, you should also step out of your VR environment entirely during these breaks. This allows your brain to recalibrate and ensures that you don't experience lingering dizziness or disorientation. Use this time to hydrate, stretch, and relax before resuming your VR activities.

2. Adjust the Display Settings

The Vive Focus Vision allows you to adjust various display settings, such as brightness and contrast, to optimize your viewing experience. For users who experience eye strain, lowering the brightness slightly or adjusting the contrast can reduce glare and help alleviate discomfort. Additionally, adjust the refresh rate to a setting that works best for you, as higher refresh rates tend to provide smoother visuals that reduce strain.

3. Monitor for Motion Sickness

Motion sickness is a common issue when using VR, and it can occur if your virtual movement doesn't align with your physical movement. For example, fast or jerky motion in VR can confuse your body's balance sensors, leading to nausea or dizziness. If you experience these symptoms, immediately stop the activity and remove the headset.

To help reduce the likelihood of motion sickness, try to avoid fast or erratic movements, especially in applications that require rapid locomotion. Some VR applications include comfort settings, such as teleportation movement or vignetting (a visual effect that narrows the field of view during motion), which can reduce the sensation of motion sickness. If available, enable these options for a more comfortable experience.

4. Use Anti-Glare Lenses or Glasses

If you wear glasses or have trouble with glare, consider using lens protectors or anti-glare lenses designed for VR. These can help improve clarity and reduce strain on the eyes. Specialized VR glasses are available that fit over your regular glasses and offer a more comfortable fit, especially if you need prescription lenses. Additionally, using a lens cleaning cloth regularly to remove dust and smudges from the lenses can help maintain sharp visuals.

Proper Posture and Movement Guidelines:

Proper posture and movement are crucial for a safe and comfortable VR experience. Misalignment can lead to neck pain, back strain, and other physical issues. Below are some tips to help maintain good posture while using your HTC Vive Focus Vision.

1. Maintain Neutral Posture

While in VR, it's essential to maintain a neutral posture to prevent strain on your neck and back. Keep your spine aligned and avoid slouching. Stand with your feet shoulder-width apart, and when sitting, ensure that your feet are flat on the floor with your knees at a 90-degree angle. Avoid tilting your head too far forward, backward, or sideways, as this can strain your neck muscles over time.

2. Move Within Your Play Area

Be mindful of your movement within the play area. While VR enables you to engage in dynamic actions, such as walking, jumping, or reaching, it's essential to stay within the designated play boundaries. Constantly shifting between standing, sitting, or kneeling positions may also lead to muscle fatigue, so make sure to take breaks and switch positions periodically. You should also use VR accessories that improve comfort, such as padded straps or headsets with adjustable padding.

3. Body Alignment During Gameplay

For users who engage in active VR experiences, such as fitness apps or games that require full-body movement, maintaining good alignment while performing exercises is essential. To avoid injury, ensure that your body is properly aligned during physical movements, particularly when stretching or lifting. Avoid overexerting yourself, and be cautious about the intensity of movement, especially if you are new to VR workouts.

How to Take Breaks and Stay Safe:

Taking breaks and practicing proper VR use habits is key to avoiding health issues such as eye strain, fatigue, or physical discomfort. Use these guidelines to stay safe while enjoying your VR sessions:

1. Set a Timer for Breaks

To help manage your VR time and ensure you take breaks regularly, set a timer. You can use your phone, a smartwatch, or the headset's built-in features to remind you to take a break. Even if you're enjoying a particularly immersive experience, it's important to step away every 30-45 minutes to rest both your eyes and your body.

2. Check for Fatigue

Always listen to your body. If you feel dizzy, lightheaded, or overly fatigued, immediately stop using the headset and rest. If you continue to experience discomfort after taking a break, it may be a sign to reduce the length of your VR sessions or adjust settings like brightness or movement.

3. Hydrate and Stretch

Hydration is key to preventing fatigue and maintaining focus during VR use. Be sure to drink plenty of water before, during, and after your VR sessions. Additionally, take time to stretch your arms, legs, and neck to relieve any tension that may have built up during use. Gentle stretches can help improve circulation and reduce muscle stiffness.

4. End Sessions Gradually

As you approach the end of your VR session, try to gradually ease out of the experience. Remove the headset, stretch, and take a few moments to readjust to the physical world. If you're playing a game, try to finish at a natural stopping point rather than abruptly cutting off your VR session.

Conclusion:

While the HTC Vive Focus Vision offers an exciting and immersive virtual experience, ensuring your safety and health is equally important. By following the VR safety tips, reducing the risk of eye strain, motion sickness, and physical discomfort, and taking regular breaks, you can enjoy long, comfortable sessions without compromising your health. Proper posture and movement are crucial for minimizing strain, and by adjusting the headset and play area to your needs, you can optimize your VR experience.

Chapter 14.

Technical Specifications

The HTC Vive Focus Vision is a high-end standalone mixed-reality headset designed for both consumer and business applications. With cutting-edge hardware and advanced features, it delivers a seamless virtual and mixed reality experience. This chapter breaks down the technical specifications of the Vive Focus Vision, providing detailed information about its display, lenses, camera systems, processing power, battery life, and supported media formats.

Display and Lens Specifications:

One of the standout features of the HTC Vive Focus Vision is its advanced display system, which provides high-resolution visuals and a wide field of view, essential for creating an immersive VR experience. Here's a breakdown of its display and lens specifications:

1. Resolution

The Vive Focus Vision features dual LCD displays, each with a resolution of 2,448 x 2,448 pixels. This high resolution ensures that the visuals are sharp, clear, and detailed, making it suitable for both immersive VR gaming and professional applications that require high visual fidelity, such as architectural visualization or medical training.

The combined pixel density of 2,448 x 2,448 per eye is well above the standard for many consumer-level VR headsets, contributing to a more lifelike and realistic experience. This also minimizes the screen-door effect, a common issue in lower-resolution displays where users can see the spaces between pixels.

2. Field of View (FOV)

The Vive Focus Vision offers a 120-degree horizontal field of view (FOV), which is crucial for a fully immersive experience. The wider the FOV, the more natural and enveloping the virtual environment appears. The 120-degree FOV also reduces the "tunnel vision" effect found in headsets with narrower views, providing a broader perspective for users.

3. Lens Technology:

The headset is equipped with Fresnel lenses that help maintain clear focus across the entire field of view. Fresnel lenses are designed to reduce optical distortions and provide a consistent visual experience, minimizing visual artifacts like glare and halo effects that are common in lower-quality lenses.

Additionally, the Vive Focus Vision allows users to adjust the interpupillary distance (IPD), enabling a more personalized and comfortable viewing experience. This adjustment helps optimize the distance between the lenses and your eyes, ensuring that the virtual world appears sharp and natural, reducing the chance of eyestrain.

Camera and Sensor Details:

The HTC Vive Focus Vision integrates a variety of cameras and sensors to enhance tracking accuracy, user interaction, and mixed-reality (MR) capabilities.

1. Tracking Cameras

The headset is equipped with four front-facing tracking cameras that allow for precise hand tracking and spatial awareness. These cameras work in conjunction with the headset's sensors to track user movements and gestures in real time. This enhances the overall experience, particularly in interactive applications, gaming, and activities where physical motion is essential.

2. Depth Sensors

In addition to the tracking cameras, the Vive Focus Vision includes depth sensors that assist with spatial awareness and environmental mapping. These sensors are essential for mixed-reality features, allowing the device to accurately sense the physical world around you. This allows for smooth interaction between the real and virtual environments, enabling features like color passthrough, where you can see the real world in full color while interacting with virtual objects.

3. Color Passthrough Cameras

The Vive Focus Vision supports color passthrough, which allows users to see their surroundings through the cameras in full color. This feature is especially important for mixed-reality (MR) applications where users need to interact with virtual objects in real-time while maintaining awareness of the physical world. The color passthrough feature is powered by dual color cameras that work in tandem with the depth sensors to provide a realistic representation of the environment.

4. Eye-Tracking Sensors

Eye-tracking is one of the key features of the Vive Focus Vision. The built-in eye-tracking sensors allow the headset to track where you're looking within the virtual environment. This data can be used for dynamic foveated rendering (discussed later) and also to create more immersive experiences by responding to the user's gaze. Eye tracking improves interaction with virtual environments and helps to reduce rendering load by focusing system resources on the areas where the user is looking.

Processor and Memory Information:

The Vive Focus Vision packs a powerful processing system designed to handle the intensive demands of virtual reality, mixed reality, and real-time tracking. Here's a breakdown of the key components:

1. Processor

The headset is powered by the Qualcomm Snapdragon XR2 Gen 1 chipset. This processor is specifically designed for augmented reality (AR) and virtual reality (VR) devices, offering a significant performance boost over previous Snapdragon models. The XR2 Gen 1 is built to handle high-resolution displays, advanced sensor data, and demanding VR applications, ensuring smooth performance with minimal latency.

The Snapdragon XR2 Gen 1 chipset supports wireless connectivity, integrated AI features, and a broad range of VR and MR content. This makes the Vive Focus Vision a powerful device for both consumers and professionals who need high-quality mixed-reality experiences.

2. Memory and Storage

The Vive Focus Vision is equipped with 12GB of RAM, which ensures smooth operation when running demanding applications, particularly those involving 3D rendering or simultaneous sensor data processing. The large amount of RAM helps prevent lag or stuttering during extended use, providing a fluid experience.

For storage, the headset includes 128GB of internal storage. This storage is ample for installing a wide range of VR applications, games, and multimedia content. Users can store games and VR experiences locally without relying on external devices, making the headset an all-in-one, standalone solution for VR and MR content.

Battery Life and Charging Time:

Battery life is a crucial consideration for any standalone VR headset, and the Vive Focus Vision delivers in this regard.

1. Battery Life

The Vive Focus Vision features a high-capacity rechargeable battery that offers up to 3-4 hours of continuous use. This is enough for extended gaming sessions, business applications, or training modules. For most users, the headset's battery life is sufficient for typical VR sessions, but for those who need longer usage times, it's recommended to take breaks and recharge between sessions.

2. Charging Time

The Vive Focus Vision has a fast-charging capability that allows it to reach a full charge in about 2-2.5 hours. The charging time can vary slightly depending on the charger and the device's current battery level. For those who are constantly on the go, the device also supports USB-C charging, providing a convenient and flexible charging option.

3. Battery Management Features

The Vive Focus Vision includes several battery management features to help extend the overall lifespan of the device's battery. These include power-saving modes that adjust the device's brightness and processing power when idle, as well as automatic shutdown after extended periods of inactivity.

Supported Video and Audio Formats:

For those looking to use the Vive Focus Vision for media consumption, such as watching VR movies, streaming content, or enjoying 360-degree videos, it supports a wide range of video and audio formats.

1. Video Formats

The headset supports popular video formats such as MP4, MKV, AVI, and MOV. It can also handle 360-degree videos, which are essential for fully immersive VR media consumption. The high resolution of the headset's display allows users to experience content in vivid detail, whether it's a movie, a concert, or a virtual tour.

2. Audio Formats

The Vive Focus Vision also supports a variety of audio formats, including MP3, WAV, FLAC, and AAC. While the headset has built-in spatial audio capabilities, users can also connect external Bluetooth headphones or use the Vive Focus Vision's USB-C port for wired audio. This flexibility ensures that users can enjoy high-quality sound tailored to their preferences.

Conclusion:

The HTC Vive Focus Vision is packed with advanced specifications that make it a top-tier choice for both consumers and professionals in the VR and MR markets. From its high-resolution display and powerful Snapdragon XR2 chipset to its comprehensive camera system and eye-tracking capabilities, the Vive Focus Vision offers an exceptional mixed-reality experience. Whether you're using it for gaming, training, or professional applications, its robust technical specifications provide a reliable and immersive platform for a wide range of activities.

Chapter 15.

Warranty and Support

When investing in a high-tech device like the HTC Vive Focus Vision, it's important to understand the warranty and support options available to you. HTC provides comprehensive warranty coverage and customer service resources to ensure that you get the most out of your VR headset. This chapter covers the details of warranty coverage, how to get technical support, the importance of registering your product, and how to contact HTC's customer service for assistance.

Warranty Coverage and Terms:

HTC offers a standard warranty for the Vive Focus Vision that covers manufacturing defects and malfunctions in materials or workmanship under normal use. Here's a breakdown of what you can expect from the warranty:

1. Warranty Period

The standard warranty for the HTC Vive Focus Vision is typically 1 year from the date of purchase. During this period, HTC will repair or replace any defective parts free of charge. This warranty applies to the headset itself, the controllers, and any included accessories like the Vive Focus Vision's facial tracker, provided these are used as intended and according to the manufacturer's guidelines.

2. What's Covered

The warranty covers defects in material and workmanship that occur under normal usage. This includes issues such as faulty components, malfunctioning hardware, and software bugs that arise due to manufacturing defects. If you experience issues with the display, lenses, buttons, sensors, or any core components of the headset, these should be covered by the warranty.

3. What's Not Covered

It's important to note that the warranty does not cover damage caused by misuse, accidental damage, or unauthorized repairs. If the headset is dropped, exposed to liquids, or modified in any way that voids the warranty, HTC will not be liable for repairs or replacements. Additionally, wear and tear, such as scratches on the lenses or damaged cables due to improper handling, is not covered.

4. Extended Warranty Options

In some regions, HTC offers extended warranty plans for additional protection beyond the standard coverage. This can be particularly useful if you plan to use the Vive Focus Vision intensively for professional purposes. You can check with HTC or authorized dealers for information about extended warranties and service plans.

Getting Technical Support:

Should you encounter any technical issues with the HTC Vive Focus Vision, HTC provides several support options to help resolve problems. Whether you're dealing with setup issues, hardware malfunctions, or troubleshooting software errors, HTC's support team can guide you through solutions.

1. Online Support Portal

The HTC support portal is a comprehensive online resource where you can find answers to frequently asked questions, troubleshooting guides, and detailed user manuals. The portal includes troubleshooting steps for common issues such as display problems, connectivity errors, and software malfunctions.

You can search for specific issues or browse categories of problems related to setup, configuration, and performance. The online support portal is the first place to visit for basic troubleshooting and technical solutions.

2. Live Chat Support

HTC also offers live chat support, allowing you to connect with a customer service representative in real-time. This service is helpful for more complex issues that may require immediate attention. Live chat representatives can guide you through step-by-step troubleshooting, offer solutions, and help you resolve problems with your headset or accessories.

3. Phone Support

For more personalized assistance, HTC provides phone support for its customers. You can contact HTC's customer service hotline, where trained support agents are available to assist with your issue. The phone support team can help with everything from software troubleshooting to coordinating warranty claims and product repairs.

Keep in mind that phone support may be available during specific hours depending on your region. Check the HTC website for phone support hours and contact information based on your country or region.

4. Community Forums

HTC also has a community forum where users can share experiences, ask questions, and provide solutions to common issues. The forum is a great place to connect with other Vive Focus Vision users, especially if you are looking for advice on specific applications or want to share tips on improving your experience. HTC moderators and community managers also monitor the forum to ensure that information shared is accurate and up-to-date.

Registering Your Product:

Registering your HTC Vive Focus Vision is an important step in ensuring that you receive the full benefits of your warranty and support options. Product registration helps HTC track your device's serial number and purchase information, making it easier to offer support if issues arise.

1. Why Register?

When you register your Vive Focus Vision, HTC can quickly verify your warranty status and provide faster customer service. Registration also ensures that you receive important product updates, software releases, and promotional offers. It's a simple way to ensure that your device is properly documented for warranty and support purposes.

2. How to Register

To register your Vive Focus Vision, you'll need to visit HTC's product registration page on their official website. You'll be asked to provide basic information such as:

The serial number of your Vive Focus Vision headset (usually found on the device or in the documentation).

Proof of purchase, such as a receipt or invoice from the retailer where you purchased the headset.

Your contact information and shipping address.

Once you complete the registration form, HTC will confirm your registration via email. It's a good idea to save this email for future reference, as it will contain a record of your registration and purchase.

3. What Happens After Registration?

After registration, you'll have access to HTC's customer service resources, including the ability to submit warranty claims, request technical support, and receive software updates. Additionally, registering your device may give you access to special offers, including discounts on accessories or extended warranty options.

HTC Customer Service Information:

HTC's customer service team is available to assist with any issues related to your Vive Focus Vision, from hardware malfunctions to software questions. Whether you need to file a warranty claim, troubleshoot a problem, or get guidance on using your headset, HTC's customer service is dedicated to providing timely support.

1. Contact Methods

Phone Support:

HTC's customer service phone lines are available in most regions, and the team is trained to handle both technical and warranty-related inquiries.

Live Chat:

The HTC website offers live chat support, where you can talk directly with a representative for immediate help.

Email Support:

You can also contact HTC's support team via email. The email address can be found on the HTC support website, and you can expect a response within 24-48 hours.

Social Media:

HTC also has active customer service accounts on social media platforms, where you can ask questions and receive updates on product news, including troubleshooting and technical support.

2. Service Centers and Repairs

If your Vive Focus Vision needs repair, HTC has authorized service centers where you can send your headset for inspection and fixing. The process for sending your device in for repair depends on the warranty terms, so be sure to check with HTC for specific details. If your headset requires parts replacement or an extended repair period, HTC will inform you of the expected timeline and provide updates on the repair status.

Conclusion:

HTC provides robust warranty coverage and support options for the Vive Focus Vision, ensuring that users have access to resources in the event of issues or malfunctions. From online support portals and live chat to phone support and community forums, HTC offers multiple ways for users to get assistance. Registering your device not only ensures that you're covered under warranty but also gives you access to the latest updates and exclusive offers. Should you need further help, HTC's customer service is ready to assist with any inquiries, making it easier for you to enjoy a seamless VR experience.

Conclusion

The HTC Vive Focus Vision stands as one of the most advanced mixed-reality headsets available today. Offering an array of features designed for both consumers and professionals, this device brings high-quality visuals, powerful tracking capabilities, and immersive experiences in both virtual and mixed-reality environments. Whether you're a gamer, a developer, or someone looking to use VR for business or creative work, the Vive Focus Vision combines cutting-edge technology with a user-friendly experience. In this conclusion, we will recap the key features of the Vive Focus Vision, offer tips for optimizing your VR experience, and provide essential contact information to help you should you need additional support.

Summary of Features:

The HTC Vive Focus Vision features a host of technologies that enhance the overall VR and MR experience. At the heart of the headset is its impressive display, which includes dual 2448 x 2448 LCD panels offering a sharp and vibrant resolution. The high refresh rate of 90Hz ensures a smooth visual experience, reducing motion blur and enhancing overall clarity in fast-moving scenes. The 120-degree horizontal field of view contributes to an expansive and natural viewing angle, making the VR environment feel more immersive.

One of the standout features of the Vive Focus Vision is its built-in eye-tracking functionality. This eye-tracking system allows for dynamic adjustment of the lenses and provides support for foveated rendering, which intelligently adjusts the resolution in the peripheral areas of the lens, offering better visual fidelity and reducing the computational load. This feature makes the experience more efficient while providing stunning visuals.

Additionally, the Vive Focus Vision boasts advanced hand and body tracking capabilities, facilitated by a combination of infrared sensors, cameras, and optional accessories such as the Vive Facial Tracker and Vive Ultimate Trackers. These accessories provide users with more granular control and help bring more natural and expressive movement into the virtual space. The body tracking functionality in particular opens up new possibilities in both gaming and professional applications, enabling users to create avatars that react in real time to their physical actions.

For those who need to use the Vive Focus Vision for PC VR, the headset supports a wired connection through a DisplayPort adapter. This allows users to access PC VR content via SteamVR and Viveport, enabling a higher-quality experience compared to standalone mode. Wireless streaming is also supported, giving users the flexibility to experience content without being tethered to a computer, although the wired connection offers superior performance and stability.

The Vive Focus Vision is also equipped with mixed-reality (MR) features, including color passthrough. The dual color cameras and depth sensors allow users to blend the virtual and real worlds, enhancing interaction with both environments. This is ideal for applications where users need to remain aware of their physical surroundings while immersed in VR.

Tips for Optimizing Your VR Experience:

To get the most out of your Vive Focus Vision, there are several steps you can take to enhance your VR experience:

1. Positioning and Comfort:

Ensure that your headset is properly adjusted for comfort. The Vive Focus Vision is a premium device, but the weight can become uncomfortable if not properly balanced. Make sure to adjust the strap for a snug fit and position the battery at the rear to help balance the weight. Additionally, adjusting the interpupillary distance (IPD) and eye relief can improve visual clarity and comfort, especially for prolonged use.

2. Use Eye Tracking to Improve Performance:

Take advantage of the eye-tracking and dynamic foveated rendering to improve both visual quality and performance. Eye-tracking allows the system to focus high-resolution rendering on the area where your eyes are looking, while peripheral areas receive lower-resolution rendering, reducing strain on the system. This leads to more detailed visuals where you need them and a smoother performance overall.

3. Take Breaks to Avoid Motion Sickness:

While VR experiences are engaging, they can also cause fatigue or motion sickness if used for extended periods. It's important to take breaks regularly. A good rule of thumb is to take a break every 20 to 30 minutes, particularly if you're playing fast-paced games or engaging in intense simulations.

4. Keep Your Lenses Clean:

The lenses on your Vive Focus Vision are essential for clarity and immersion. Make sure to clean them regularly with a microfiber cloth to prevent dust, smudges, or fingerprints from obscuring your view. Always store your headset in a case or protective bag to avoid scratches.

5. Optimize Your Play Space:

Ensure you have enough space to move freely without running into obstacles. Clear a 6.5-foot-by-6.5-foot area to get the most out of room-scale experiences. The Vive Focus Vision's tracking cameras are designed to track your movement accurately, but a clean, open play area will ensure optimal performance.

6. Update Firmware Regularly:

HTC provides regular firmware and software updates to ensure that your Vive Focus Vision stays current with the latest features and improvements. Keep an eye on the updates and install them as soon as they're available to enhance your overall experience and fix any known bugs.

7. Explore Mixed Reality:

Experiment with the color passthrough and mixed-reality features to create a seamless interaction between the real and virtual worlds. This is particularly useful for applications that require you to move around or interact with real-world objects while still being immersed in a virtual environment.

8. Use High-Performance Hardware for PC VR:

If you plan on using the Vive Focus Vision for PC VR, ensure that your PC meets the required specifications to deliver optimal performance. A powerful GPU, such as an RTX 30-series or 40-series card, will help deliver the best possible visuals and minimize latency during gameplay.

Contact Information for Additional Help:

While the Vive Focus Vision is designed to be user-friendly, there may still be instances where you require technical assistance or additional support. HTC provides several avenues for customers to get the help they need.

HTC Support Website:

The official HTC support site is a great first stop for finding troubleshooting guides, FAQs, and user manuals. You can access the website at www.vive.com/support to find helpful articles on setup, usage, and troubleshooting.

Customer Service:

If you need more personalized assistance, HTC offers customer service via email and phone. You can reach their customer support team by calling their hotline or sending an email through the website. Be sure to have your Vive Focus Vision's serial number handy when contacting support to expedite the process.

Community Forums:

The HTC Vive forums are another excellent resource for users to share their experiences, ask questions, and find solutions to common problems. Many issues have already been discussed in the forums, so browsing through them might provide quick answers.

Warranty and Returns:

If you experience a hardware issue that requires a replacement or repair, HTC's warranty policy covers most defects. Be sure to review the warranty terms, which can be found in your device's manual or on the HTC website. You can initiate a warranty claim through the support page, where you'll be guided on how to return or replace defective products.

In conclusion, the HTC Vive Focus Vision offers a remarkable combination of high-end VR features, including excellent visuals, tracking, and mixed-reality capabilities. By following the tips provided and utilizing the available resources, you can optimize your experience to get the most out of your device. Should you need further assistance, HTC's customer service and support channels are always ready to help. Enjoy your immersive journey into the world of VR and mixed reality with the HTC Vive Focus Vision.

www.ingramcontent.com/pod-product-compliance
Lightning Source LLC
Chambersburg PA
CBHW071027240526
45469CB00006BD/2118